H.W. Watson

A treatise on the application of generalised coordinates to the kinetics of a material system

H.W. Watson

A treatise on the application of generalised coordinates to the kinetics of a material system

ISBN/EAN: 9783742892287

Manufactured in Europe, USA, Canada, Australia, Japa

Cover: Foto ©berggeist007 / pixelio.de

Manufactured and distributed by brebook publishing software (www.brebook.com)

H.W. Watson

A treatise on the application of generalised coordinates to the kinetics of a material system

Clarendon Press Series

A TREATISE

ON THE

APPLICATION OF GENERALISED COORDINATES

TO THE

KINETICS OF A MATERIAL SYSTEM

BY

H. W. WATSON, M.A.
FORMERLY FELLOW OF TRINITY COLLEGE, CAMBRIDGE

AND

S. H. BURBURY, M.A.
FORMERLY FELLOW OF ST. JOHN'S COLLEGE, CAMBRIDGE

Oxford
AT THE CLARENDON PRESS
1879

PREFACE.

THE treatment of the kinetics of a material system by the method of generalised coordinates was first introduced by Lagrange, and has since his time been greatly developed by the investigations of different mathematicians.

Independently of the highly interesting, although purely abstract science of theoretical dynamics which has resulted from these investigations, they have proved of great and continually increasing value in the application of mechanics to thermal, electrical and chemical theories, and the whole range of molecular physics.

The object of the following short treatise is to conduct the student to the most important results hitherto obtained in this subject, by demonstrations free from intricate analysis and based, as far as possible, upon the direct application of mechanical and geometrical considerations.

The earlier propositions contain, for the most part, little that is absolutely original so far as results are concerned, but in the concluding portion of the work the theory of *Least Action* and *Kinetic Foci* has been investigated from a somewhat novel point of view, and in a manner which it is hoped may tend to throw some additional light upon this obscure and difficult subject.

The language and notation of Quaternions have been employed in two or three instances, but never to such an extent as to break the continuity of the treatise or to prove a hindrance to the student who is unacquainted with that branch of mathematics.

TABLE OF CONTENTS.

CHAPTER IV.

CHAPTER V.

CHAPTER I.

DEFINITIONS AND GENERAL THEOREMS.

Generalised Coordinates.

ARTICLE 1.] WHEN the position of every point of a material system can be determined in terms of any independent variables n in number, the system is said to possess *n degrees of freedom*, and the n independent variables are called the *generalised coordinates.*

The choice of the particular independent variables is perfectly arbitrary, and may be varied indefinitely, but the number of degrees of freedom cannot be either increased or diminished.

In a rigid body free to move in any manner there are six degrees of freedom, and the generalised coordinates most frequently chosen in this case are the three rectangular coordinates of some point in the body and three angular coordinates determining the orientation of the body about that point, generally the angles θ, ϕ, ψ of ordinary occurrence in rigid dynamical problems.

When the body degenerates into a material straight line the number of degrees of freedom is reduced to five; and when this straight line is constrained to move parallel to some fixed plane the number of degrees of freedom is still further reduced to four.

A chain of n links, in which each link is a material straight line, has in the most general case $2n+3$ degrees of freedom, and if one point in this chain be fixed the number is reduced to $2n$, and we might choose for our generalised coordinates the $2n$ angles which determine the directions of the links.

And so on for many other examples. The n coordinates are very generally denoted by $q_1, q_2, \ldots q_n$.

Generalised Components of Momentum.

2.] The complete knowledge of the state of any material system embraces not only its configuration but its motion at any instant.

Suppose the velocity of each element of the system to be known, and let it be multiplied by the mass of that element so as to obtain the momentum of the element, and let the infinitesimal variation δq_r be given to any coordinate q_r.

Then, if f be the momentum of the element, and if l be the distance of the projection of that element upon some fixed line to which the velocity of the element is instantaneously parallel, measured from some fixed point in that line, it follows from definition that l is some known function of the q's, and the virtual moment of the momentum of the element consequent on the variation δq_r is clearly $f\frac{dl}{dq_r}\delta q_r$, and the sum of such virtual moments for the whole system is $\left(\Sigma f\frac{dl}{dq_r}\right)\delta q_r$.

The coefficient $\Sigma\left(f\frac{dl}{dq_r}\right)$ is called *the generalised component of momentum corresponding to the coordinate* q_r.

In the actual motion each element, as m, is describing a determinate curve such that the length s of that curve measured from a fixed point in it is a known function of the q's, and the velocity of m is $\frac{ds}{dt}$, therefore the generalised component of momentum of the system corresponding to the coordinate q_r is $\Sigma m\frac{ds}{dt}\cdot\frac{ds}{dq_r}$.

If x, y, z be the rectangular coordinates of m, this generalised component of momentum may be written

$$\Sigma m\left(\frac{dx}{dt}\frac{dx}{dq_r}+\frac{dy}{dt}\frac{dy}{dq_r}+\frac{dz}{dt}\frac{dz}{dq_r}\right).$$

In the language of quaternions, ρ, the vector from the origin to any element of the system of mass m, may be regarded as a function of the n scalar variables $q_1, \dots q_n$, and we then have, if the above-mentioned component of momentum be denoted by p_r,

$$p_r = -\Sigma m S\left(\frac{d\rho}{dt}\frac{d\rho}{dq_r}\right).$$

A similar definition applies to the *generalised component of effective force* of the system corresponding to the coordinate q_r, viz. the sum of the virtual moments of the effective forces of all the particles corresponding to the variation δq_r; and from reasoning exactly similar to the above it follows that this generalised component may be written

$$\Sigma m\left(\frac{d^2x}{dt^2}\frac{dx}{dq_r} + \frac{d^2y}{dt^2}\frac{dy}{dq_r} + \frac{d^2z}{dt^2}\frac{dz}{dq_r}\right);$$

or again using the notation of quaternions,

$$-\Sigma m S\left(\frac{d^2\rho}{dt^2}\frac{d\rho}{dq_r}\right).$$

The following notation will be generally employed in dealing with generalised coordinates:—

(1) The coordinates will be denoted by $q_1, q_2 \dots q_n$, as above stated.

(2) The differential coefficients of these coordinates with regard to the time t, also called *the generalised components of velocity*, will be denoted by $\dot{q}_1, \dot{q}_2 \dots \dot{q}_n$.

(3) The generalised components of momentum will be denoted by $p_1, p_2 \dots p_n$.

Generalised Components of Force.

3.] Let the material system be acted on by any given forces, and let F be one of these forces.

Then if l be the distance of the projection of the point of application of F upon some fixed straight line parallel to the direction of F, measured from some fixed point in that line, it follows from definition that l is some known function of the q's, and the virtual moment of F consequent upon any infinitely small variation δq_r in the coordinate q_r is $F\frac{dl}{dq_r}\delta q_r$, and the sum of such virtual moments for all the forces acting on the system is $\Sigma\left(F\frac{dl}{dq_r}\right)\delta q_r$.

The coefficient of δq_r or $\Sigma\left(F\frac{dl}{dq_r}\right)$ is called *the generalised component of force corresponding to the coordinate* q_r.

If the coordinates of the point of application of F referred to any fixed rectangular axes be x, y, and z, and if the corresponding components of F be X, Y, and Z, this generalised component of force becomes

$$\Sigma\left(X\frac{dx}{dq_r}+Y\frac{dy}{dq_r}+Z\frac{dz}{dq_r}\right).$$

Generalised Components of Impulse.

4.] When the forces in action are very large, and the time during which they act is very short, they are called impulses, and are generally measured by the time integrals of the forces.

If F be any impulse measured in the manner just described, and if X, Y, Z be its rectangular components, and if l, x, y, z have the same meanings as in the last article, then the generalised component of impulse corresponding to the coordinate q_r will be

$$\Sigma F\frac{dl}{dq_r} \text{ or } \Sigma\left(X\frac{dx}{dq_r}+Y\frac{dy}{dq_r}+Z\frac{dz}{dq_r}\right).$$

It must be carefully remembered that the virtual moment of an impulse does not, as in the case of finite forces, represent work done consequent on the variation δq_r.

When an impulse is spoken of as the time integral of a force, it is only in a particular case that the term is used with strict accuracy, namely when the direction of the infinitely large force which acting for an infinitely short time produces the impulse remains the same during that short time.

In such a case, if P be the large force, F the impulse, and τ the short time of action, F is accurately equal to $\int_0^\tau P dt$. But it is quite conceivable that P, the constituent force of the impulse, should vary in direction as well as intensity during the time τ. In this case we cannot say that $F=\int_0^\tau P dt$, but we must say that F is the resultant of all the momenta $P_1 dt_1$, $P_2 dt_2$, &c. added throughout that interval.

In the former case, where the force is fixed in direction during the time τ, we obtain the generalised component of impulse, as we have said above, by writing $\int_0^\tau P\,dt$ for F in the expression $\Sigma F\frac{dl}{dq}$.

In the latter we can only obtain this generalised component by resolving P in fixed directions during the time of its action, and thus we are restricted to the expression

$$\Sigma\left(\int_0^\tau X dt\,\frac{dx}{dq_r} + \int_0^\tau Y dt\,\frac{dy}{dq_r} + \int_0^\tau Z\,dt\,\frac{dz}{dq_r}\right),$$

X, Y, and Z being the rectangular components of P at any instant.

5.] The terms generalised components of momentum, force, impulse, are very convenient for use, but it is important to remember that they are frequently only names, and do not represent actual forces, impulses, or momenta, still more rarely are they component forces, impulses, or momenta in the ordinary meaning of the term, i.e. such that their simultaneous action or existence is equivalent to the forces, impulses, or momenta acting on, or existing in, the system.

For example, let the system of impressed forces be two parallel forces F and $-F$ at right angles to an axis and at the distances $a+b$ and b from it, and let one of the coordinates be the angle θ between a plane fixed in the body containing that axis and a similar plane fixed in space.

The virtual moments of the forces consequent on the small variation $\delta\theta$ are $F(a+b)\,\delta\theta$ and $-Fb\,.\,\delta\theta$, and the generalised component of force corresponding to θ is Fa, that is to say, it is a couple and not a force.

Or again, suppose we are considering the case of a single material particle referred to axes Ox and Oy at the angle a, and acted on by impressed forces parallel to these axes equal to X and Y respectively.

If the coordinate x be varied by δx, the virtual velocities of

X and Y will be δx and $\delta x \cos a$ respectively, and the sum of the virtual moments of X and Y will be $(X + Y\cos a)\delta x$, so that the generalised component of force corresponding to x is $X + Y\cos a$, and similarly that corresponding to y is $Y + X\cos a$; the generalised components are therefore, in this case, forces but not component forces.

Again, suppose that the motion of a particle m is referred to the last-mentioned axes, and that the velocities parallel to x and y are u and v respectively. Then it will follow by similar reasoning that the generalised components of momentum are $m(u + v\cos a)$ and $m(v + u\cos a)$ respectively, i. e. they are not component momenta in the ordinary acceptation of that term.

A very interesting example of generalised components of momentum is afforded by the case of a rigid body moving about a fixed point and referred to the ordinary angular coordinates θ, ϕ, ψ.

Let A, B, and C be the principal moments of inertia about the fixed point, and ω_1, ω_2, ω_3 the angular velocities about the principal axes. To find the generalised components of momentum corresponding to θ, ϕ, ψ, we must resolve the couples of momenta $A\omega_1$, $B\omega_2$, $C\omega_3$ into three pairs, so that one out of each pair has its axis coincident with the axis of θ, ϕ, or ψ as the case may be, and the remaining couple of each pair has its axis perpendicular to the axis of θ, ϕ, or ψ. Then neglecting the second couple of each pair we have the required generalised component of momentum.

For the coordinate θ this is found to be by obvious resolution

$$B\omega_2 \cos\phi + A\omega_1 \sin\phi.$$

For ϕ it is $\quad + C\omega_3$.

For ψ it is $C\omega_3 \cos\theta + (B\omega_2 \sin\phi - A\omega_1 \cos\phi)\sin\theta$.

6.] Any material system may be regarded as a collection of discrete particles whose positions are constrained to satisfy certain geometrical conditions by means of internal forces acting amongst themselves.

D'Alembert's principle asserts that when its effective force reversed is applied to each particle, there will be equilibrium among the impressed and reversed effective forces throughout the system, and that the internal forces above mentioned are in equilibrium among themselves.

That is to say, that if the system be slightly displaced, with due regard to the geometrical conditions, the sum of the virtual moments of the internal forces taken throughout the system will be zero, and may be neglected in forming the equation of virtual velocities. If F_q and E_q be the generalised components of impressed and effective force respectively corresponding to the coordinate q, D'Alembert's principle asserts that

$$\Sigma F_q \delta q = \Sigma E_q \delta q,$$

the summation being for all the q's.

And in like manner, if P_q be the generalised component of impulse acting on a system at rest, corresponding to the coordinate q, and if p_q be the corresponding component of momentum in the motion caused by the impulse, D'Alembert's principle asserts that $$\Sigma P_q \delta q = \Sigma p_q \delta q.$$

Hence, since the δq's are independent, we know that the generalised component of effective force corresponding to any coordinate is equal to the corresponding generalised component of the impressed forces acting on the system, and also that the generalised component of momentum corresponding to any coordinate is equal to the corresponding generalised component of the impulses by which the actual motion might be produced in the system previously at rest.*

The Kinetic Energy.

7.] If the mass of each particle of a material system be multiplied by the square of its velocity, one half the sum of the products thus formed, taken for the whole system, is called *the kinetic energy* of the system, and is generally denoted by the

* When the system is subjected to any constraints we may either regard it as a new system altogether, with a fresh set of generalised coordinates fewer in number than before, or we may regard it as being still the same system but acted on by additional constraining forces, such that the sum of the virtual moments of these forces vanishes when the displacements are effected with due regard to the additional constraints. In this case the equations

$$\Sigma E \delta q = \Sigma F \delta q \quad \text{and} \quad \Sigma p \delta q = \Sigma P \delta q$$

will no longer be true for *all* values of the δq's but only for such values as are consistent with the constraints so imposed upon the system.

symbol T. When the generalised coordinates are geometrical magnitudes, lines, angles, and the like, this quantity T may always be expressed as a homogeneous quadratic function of the component velocities $\dot{q}_1, \dot{q}_2$, &c., with coefficients which are known functions of the q's.

For each element, as m, describes a determinate path such that s, the length of that path from some fixed point in it, is a known function of the q's;

$$\therefore \quad \frac{ds}{dt} = \frac{ds}{dq_1}\dot{q}_1 + \frac{ds}{dq_2}\dot{q}_2 + \dots + \frac{ds}{dq_n}\dot{q}_n,$$

where the coefficients $\frac{ds}{dq_1}, \frac{ds}{dq_2}$, &c. are known functions of the q's and are independent of the $\dot{q}$'s.

Hence $\left(\frac{ds}{dt}\right)^2$, and therefore $\frac{1}{2}\Sigma m\left(\frac{ds}{dt}\right)^2$ or T, must be a quadratic function of the $\dot{q}$'s with coefficients known functions of the q's.

Again, we have seen that

$$p_r = \Sigma m \frac{ds}{dt}\frac{ds}{dq_r},$$

and if the value last given of $\frac{ds}{dt}$ be substituted in this equation, it follows that p_r is a homogeneous linear function of the $\dot{q}$'s with coefficients known functions of the q's; and the same being true of each of the p's, it follows that each of the $\dot{q}$'s is a homogeneous linear function of the p's with coefficients known functions of the q's.

Since T is a homogeneous quadratic function of the $\dot{q}$'s, and since each $\dot{q}$ is a homogeneous linear function of each p, it follows that T may be expressed as a homogeneous quadratic function of the p's with coefficients known functions of the q's.

When T is thus expressed in terms of the p's it is usually written T_p, and when in terms of the $\dot{q}$'s it is written $T_{\dot{q}}$.

It may, however, happen that the equations by which the configuration of a system at any instant is determined contain the time explicitly. In such cases the time itself, t, may be taken as one of the generalised coordinates.

Or it may happen that these equations contain not only the magnitudes q_1, q_2, &c., but also the velocities $\dot{q}_1$, $\dot{q}_2$, &c.

The case in which the time t is one of the generalised coordinates may be illustrated by two particles connected by a rod which expands uniformly, or according to any other known law of time, under the influence of heat.

The case in which the $\dot{q}$'s occur as generalised coordinates may present themselves in problems dealing with rough surfaces rolling one upon another, in which the equations expressing the equality of the velocities of the points of contact cannot be readily integrated.

In all these cases, as in the simplest case first mentioned, the following equations will remain true:

$$(1) \qquad T = \tfrac{1}{2}\Sigma m\left(\frac{ds}{dt}\right)^2.$$

$$(2) \qquad \frac{ds}{dt} = \frac{ds}{dq_1}\dot{q}_1 + \frac{ds}{dq_2}\dot{q}_2 + \dots + \frac{ds}{dq_n}\dot{q}_n.$$

$$(3) \qquad p_r = \Sigma m \frac{ds}{dt}\frac{ds}{dq_r}.$$

But inasmuch as in the case of the time entering as one of the coordinates the corresponding $\dot{q}$ becomes unity, and in the case of any of the component velocities $\dot{q}_1$, $\dot{q}_2$, &c. so entering the coefficients of the type $\frac{ds}{dq}$ are not all of them independent of the $\dot{q}$'s, it will no longer be true that T may be expressed as a quadratic function either of the $\dot{q}$'s or p's.

The notations $T_{\dot{q}}$ and T_p are sometimes employed in the case where the time enters into the connecting equations; in these cases they are not quadratic functions as above, but they indicate the value of the kinetic energy expressed in terms of the coordinates and $\dot{q}$ or p respectively.*

In what follows, *where the contrary is not expressly men-*

* If the time (t) were expressed by any symbol as q_t in the connecting equations, and the kinetic energy found on the understanding that $\dot{q}_t$ was to be replaced by unity, then T would, before such evaluation of $\dot{q}_t$, be a quadratic function of all the component velocities $\dot{q}_1$, $\dot{q}_2$, ... $\dot{q}_n$ and $\dot{q}_t$.

The statement in the text refers, of course, to the expression for T in the ordinary form, i. e. after the evaluation of $\dot{q}_t$.

tioned, it is to be understood that neither the time nor any of the component velocities enter into the geometrical equations of connection of the system.

8.] In order to obtain actual expressions for the generalised components of momentum in terms of the velocities and coordinates in any particular case, it is generally most convenient to employ the following formulae. It is proved above that

$$p_r = \Sigma m \left\{ \frac{dx}{dt}\frac{dx}{dq_r} + \frac{dy}{dt}\frac{dy}{dq_r} + \frac{dz}{dt}\frac{dz}{dq_r} \right\}.$$

Also
$$\frac{dx}{dt} = \frac{dx}{dq_1}\dot{q}_1 + \frac{dx}{dq_2}\dot{q}_2 + \dots + \frac{dx}{dq_n}\dot{q}_n.$$

And similar expressions hold for $\frac{dy}{dt}$ and $\frac{dz}{dt}$.

Hence by substitution,

$$p_r = \Sigma m \left\{ \frac{dx}{dq_1}\frac{dx}{dq_r} + \frac{dy}{dq_1}\frac{dy}{dq_r} + \frac{dz}{dq_1}\frac{dz}{dq_r} \right\} \dot{q}_1$$
$$+ \quad - \quad - \quad - \quad - \quad - \quad -$$
$$+ \Sigma m \left\{ \left(\frac{dx}{dq_r}\right)^2 + \left(\frac{dy}{dq_r}\right)^2 + \left(\frac{dz}{dq_r}\right)^2 \right\} \dot{q}_r$$
$$+ \quad - \quad - \quad - \quad - \quad -$$

Example 1. An inextensible string passes over a fixed pulley A. To one end is attached a weight m_1, to the other a moveable pulley C_1. Over the moveable pulley passes another inextensible string having at its ends weights m_2 and m_3. The pulleys and string are supposed to be of inappreciable mass. If the strings hang vertically where not in contact with the pulleys, the system has two degrees of freedom, and we may take for generalised coordinates q_1, the length of the first string from the vertex of A to m_1, q_2 the length of the second string from the vertex of C_1 to m_2.

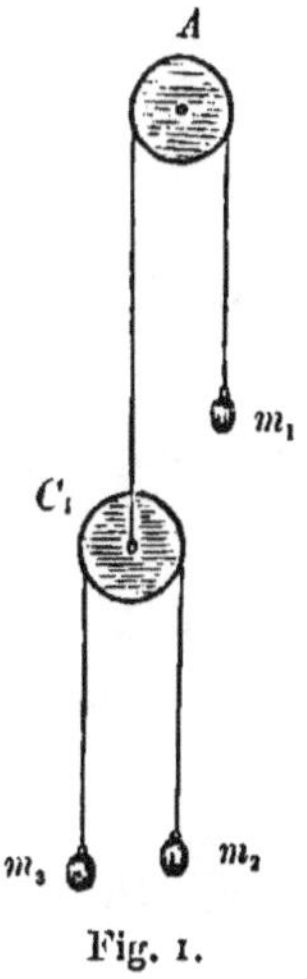

Fig. 1.

If then x_1, x_2, x_3 be the heights of m_1, m_2, and m_3 respectively above a fixed plane, we have

$$\frac{dx_1}{dq_1} = -1, \quad \frac{dx_2}{dq_1} = 1, \quad \frac{dx_1}{dq_2} = 0, \quad \frac{dx_2}{dq_2} = -1,$$

$$\frac{dx_3}{dq_1} = 1, \quad \frac{dx_3}{dq_2} = 1.$$

Therefore

$$p_1 = \Sigma m \left\{ \left(\frac{dx}{dq_1}\right)^2 \dot{q}_1 + \frac{dx}{dq_1}\frac{dx}{dq_2}\dot{q}_2 \right\}$$

$$= (m_1 + m_2 + m_3)\dot{q}_1 + (m_3 - m_2)\dot{q}_2,$$

and $$p_2 = (m_3 - m_2)\dot{q}_1 + (m_3 + m_2)\dot{q}_2.$$

If for the weight m_3 we substitute another moveable pulley C_2 over which passes another string supporting a weight m_3 and a third moveable pulley C_3, and so on till there be $\lambda - 1$ moveable pulleys, the last supporting two weights m_λ and $m_{\lambda+1}$, we shall clearly obtain the following relations

$$x_r = q_1 + \dots + q_{r-1} - q_r + b$$

for all values of r from 1 to λ inclusive, and

$$x_{\lambda+1} = q_1 + \dots + q_\lambda + c;$$

where b and c are constants.

Therefore $\frac{dx_r}{dq_s} = 1, -1$, or 0, according as r is greater, equal to, or less than s, for all values of r from 1 to λ inclusive, and $\frac{dx_{\lambda+1}}{dq_s} = 1$ for all values of s from 1 to λ, and therefore

$$p_1 = (\Sigma_1^{\lambda+1} m)\dot{q}_1 + (\Sigma_3^{\lambda+1} m - m_2)\dot{q}_2 + (\Sigma_4^{\lambda+1} m - m_3)\dot{q}_3 + \dots + (m_{\lambda+1} - m_\lambda)\dot{q}_\lambda,$$

$$p_2 = (\Sigma_3^{\lambda+1} m - m_2)\dot{q}_1 + (\Sigma_2^{\lambda+1} m)\dot{q}_2 + (\Sigma_4^{\lambda+1} m - m_3)\dot{q}_3 + \dots + (m_{\lambda+1} - m_\lambda)\dot{q}_\lambda,$$

. . . .

Example 2. Motion of a chain of λ equal uniform links each of length a in one plane, moveable in its own plane, and having one end O fixed.

This system has λ degrees of freedom, and we may take for generalised coordinates $\theta_1, \dots \theta_\lambda$, the angles made with the axis of x by the successive links beginning from O.

If r be the distance of an element of the nth link from the end of that link adjoining the next preceding, then for such element,

$$x = \Sigma_1^{n-1} a \cos\theta + r\cos\theta_n, \qquad y = \Sigma_1^{n-1} a\sin\theta + r\sin\theta_n;$$

therefore for an element of the first link,

$$\frac{dx}{d\theta_1} = -r\sin\theta_1, \qquad \frac{dy}{d\theta_1} = r\cos\theta_1;$$

$$\frac{dx}{d\theta_2} = 0, \qquad \frac{dy}{d\theta_2} = 0, \text{ \&c.}$$

For an element of link m,

$$\frac{dx}{d\theta_n} = -a\sin\theta_n, \quad \text{or} \quad -r\sin\theta_n, \quad \text{or} \quad 0,$$

according as m is greater, equal to, or less than n. Similarly,

$$\frac{dy}{d\theta_n} \text{ is } a\cos\theta_n, \quad r\cos\theta_n, \quad \text{or} \quad 0;$$

then if p_{θ_1} be the component of momentum corresponding to θ_1, we have

$$\begin{aligned} p_{\theta_1} &= \Sigma m\left(\left(\frac{dx}{d\theta_1}\right)^2 + \left(\frac{dy}{d\theta_1}\right)^2\right)\dot\theta_1 + \Sigma m\left(\frac{dx}{d\theta_1}\frac{dx}{d\theta_2} + \frac{dy}{d\theta_1}\frac{dy}{d\theta_2}\right)\dot\theta_2 \\ &\quad + \text{ - - - - - - - - - -} \\ &= (\lambda - 1 + \tfrac{1}{3}) a^3 \dot\theta_1 \\ &\qquad + (\lambda - 2 + \tfrac{1}{2}) a^3 \cos(\theta_1 - \theta_2)\dot\theta_2 \\ &\qquad + \text{ - - - - -} \\ &\qquad + \tfrac{1}{2} a^3 \cos(\theta_1 - \theta_\lambda)\dot\theta_\lambda; \\ p_{\theta_2} &= (\lambda - 2 + \tfrac{1}{2}) a^3 \cos(\theta_1 - \theta_2)\dot\theta_1 \\ &\qquad + (\lambda - 2 + \tfrac{1}{3}) a^3 \dot\theta_2 \\ &\qquad + (\lambda - 3 + \tfrac{1}{2}) a^3 \cos(\theta_2 - \theta_3)\dot\theta_3 \\ &\qquad + \text{ - - - - -} \end{aligned}$$

and $p_{\theta_\lambda} = \frac{1}{3} a^3 \dot\theta_\lambda$.

Example 3. The motion of a heavy tube in the form of a plane curve moveable in its own plane, and of a particle of mass m moveable in the tube.

This system has four degrees of freedom. Let us take for generalised coordinates,

x_1, y_1 the rectangular coordinates of a point A fixed in the tube;

θ the inclination of the tangent at A to the axis of x;

ϕ_1 the angle made by the radius vector from A to the particle with the tangent at A.

Let r be the radius vector from A to a point in the tube;

r_1 be the radius vector from A to the particle;

ϕ the angle made by the radius vector to a point in the tube with the tangent at A;

then $r = f(\phi)$ is the polar equation to the curve referred to A as pole, from which r and $\frac{ds}{d\phi}$ are known in terms of ϕ.

If x, y be the rectangular coordinates of an element of the tube,
$$x = x_1 + r\cos(\theta+\phi), \qquad y = y_1 + r\sin(\theta+\phi);$$
and in like manner for the particle,
$$x = x_1 + r_1\cos(\theta+\phi_1), \qquad y = y_1 + r_1\sin(\theta+\phi_1);$$
then for an element of the tube,
$$\frac{dx}{dx_1} = 1, \quad \frac{dy}{dx_1} = 0, \qquad \frac{dx}{dy_1} = 0, \quad \frac{dy}{dy_1} = 1;$$
$$\frac{dx}{d\theta} = -r\sin(\theta+\phi), \qquad \frac{dy}{d\theta} = r\cos(\theta+\phi);$$
$$\frac{dx}{d\phi_1} = 0, \qquad \frac{dy}{d\phi_1} = 0.$$
For the particle,
$$\frac{dx}{dx_1} = 1, \quad \frac{dy}{dx_1} = 0, \qquad \frac{dx}{dy_1} = 0, \quad \frac{dy}{dy_1} = 1;$$
$$\frac{dx}{d\theta} = -r_1\sin(\theta+\phi_1), \qquad \frac{dy}{d\theta} = r_1\cos(\theta+\phi_1);$$
$$\frac{dx}{d\phi_1} = -r_1\sin(\theta+\phi_1), \qquad \frac{dy}{d\phi_1} = r_1\cos(\theta+\phi_1).$$
Therefore if μ be the mass of an element of the tube,
$$p_{x_1} = \left(\int \mu\frac{ds}{d\phi}d\phi + m\right)\dot{x}_1$$
$$-\left\{\int \mu r\sin(\theta+\phi)\frac{ds}{d\phi}d\phi + mr_1\sin(\theta+\phi_1)\right\}\dot{\theta}$$
$$- mr_1\sin(\theta+\phi_1)\,\dot{\phi}_1;$$

$$p_{y_1} = \left(\int \mu \frac{ds}{d\phi} d\phi + m\right) \dot{y}_1$$

$$+ \left\{\int \mu r \cos(\theta+\phi) \frac{ds}{d\phi} d\phi + m r_1 \cos(\theta+\phi_1)\right\} \dot{\theta}$$

$$+ m r_1 \cos(\theta+\phi)_1 \dot{\phi}_1 ;$$

$$p_\theta = -\left\{\int \mu r \sin(\theta+\phi) \frac{ds}{d\phi} d\phi + m r_1 \sin(\theta+\phi_1)\right\} \dot{x}_1$$

$$+ \left\{\int \mu r \cos(\theta+\phi) \frac{ds}{d\phi} d\phi + m r_1 \cos(\theta+\phi)\right\} \dot{y}_1$$

$$+ \left\{\int \mu r^2 \frac{ds}{d\phi} d\phi + m r_1^2\right\} \dot{\theta} + m r_1^2 \dot{\phi}_1 ;$$

$$p_\phi = -m r_1 \sin(\theta+\phi_1) \dot{x}_1 + m r_1 \cos(\theta+\phi_1) \dot{y}_1 + m r_1^2 \dot{\theta} + m r_1^2 \dot{\phi}_1 .$$

The integrations are of course from end to end of the tube.

9.] We now proceed to prove certain propositions easily deduced from the foregoing definitions.

Proposition I. $\dfrac{dT_{\dot{q}}}{d\dot{q}_r} = p_r$.

Since $$2T = \Sigma m \left(\frac{ds}{dt}\right)^2 = \Sigma m v^2 ;$$

$$\therefore \quad \delta T = \Sigma m v \delta \frac{ds}{dt}.$$

But $$\frac{ds}{dt} = \frac{ds}{dq_1} \dot{q}_1 + \frac{ds}{dq_2} \dot{q}_2 + \dots ;$$

$$\therefore \quad \delta \frac{ds}{dt} = \frac{ds}{dq_1} \delta \dot{q}_1 + \frac{ds}{dq_2} \delta \dot{q}_2 + \dots\dots .$$

since the coefficients $\dfrac{ds}{dq_1}$, &c. are independent of the $\dot{q}$'s and the q's remain constant.

If now $\dot{q}_r$ alone varies, the remaining $\dot{q}$'s being constant, δT becomes $\dfrac{dT_{\dot{q}}}{d\dot{q}_r} \delta \dot{q}_r$;

$$\therefore \quad \frac{dT}{d\dot{q}_r} \delta \dot{q}_r = \Sigma m v \frac{ds}{dq_r} \delta \dot{q}_r,$$

or $$\frac{dT_{\dot{q}}}{d\dot{q}_r} = \Sigma m v \frac{ds}{dq_r} = p_r \text{ by definition.}$$

And this proposition is true whether the time enters explicitly into the connecting equations or not. Remembering that $T_{\dot{q}}$ is in such a case no longer a homogeneous quadratic function of the $\dot{q}$'s (see Art. 7), but is the expression for the kinetic energy in terms of the q's and $\dot{q}$'s.

PROPOSITION II. $2T = p_1\dot{q}_1 + p_2\dot{q}_2 + \&c. + p_n\dot{q}_n$.

Since $T_{\dot{q}}$ is a homogeneous quadratic function of the $\dot{q}$'s, it follows that

$$2T_{\dot{q}} = \dot{q}_1\frac{dT_{\dot{q}}}{d\dot{q}_1} + \dot{q}_2\frac{dT_{\dot{q}}}{d\dot{q}_2} + \&c. + \dot{q}_n\frac{dT_{\dot{q}}}{d\dot{q}_n}.$$

But
$$\frac{dT_{\dot{q}}}{d\dot{q}_1} = p_1, \quad \frac{dT_{\dot{q}}}{d\dot{q}_2} = p_2, \quad \&c.;$$

$$\therefore \quad 2T = p_1\dot{q}_1 + p_2\dot{q}_2 + \&c. + p_n\dot{q}_n.$$

When T is expressed in terms of the p's and $\dot{q}$'s, as in this proposition, it is written $T_{p\dot{q}}$.

PROPOSITION III. $\dfrac{dT_p}{dp_r} = \dot{q}_r$, and $\dfrac{dT_p}{dq_r} + \dfrac{dT_{\dot{q}}}{dq_r} = 0.$*

Since $T_{p\dot{q}}$, T_p, and $T_{\dot{q}}$ are three different expressions for the same magnitude, and $T_{p\dot{q}} = \frac{1}{2}\Sigma p\dot{q}$,

$$\therefore \quad T_p + T_{\dot{q}} = \Sigma p\dot{q}.$$

Let all the variables p, $\dot{q}$, and q be varied, then

$$\delta T_p + \delta T_{\dot{q}} = \delta\Sigma p\dot{q};$$

$$\therefore \quad \Sigma\frac{dT_p}{dp}\delta p + \Sigma\frac{dT_p}{dq}\delta q + \Sigma\frac{dT_{\dot{q}}}{d\dot{q}}\delta\dot{q} + \Sigma\frac{dT_{\dot{q}}}{dq}\delta q = \Sigma(p\delta\dot{q} + \dot{q}\delta p).$$

But by Proposition I
$$\frac{dT_{\dot{q}}}{d\dot{q}_r} = p_r;$$

$$\therefore \quad \Sigma\frac{dT_{\dot{q}}}{d\dot{q}}\delta\dot{q} = \Sigma p\delta\dot{q};$$

$$\therefore \quad \Sigma\frac{dT_p}{dp}\delta p + \Sigma\left(\frac{dT_p}{dq} + \frac{dT_{\dot{q}}}{dq}\right)\delta q = \Sigma\dot{q}\delta p.$$

Now the $2n$ magnitudes $p_1 \dots p_n$ and $q_1 \dots q_n$ are independent, and therefore their variations $\delta p_1 \dots \delta p_n$ and $\delta q_1 \dots \delta q_n$ are independent;

$$\therefore \quad \frac{dT_p}{dp_r} = \dot{q}_r \quad \text{and} \quad \frac{dT_p}{dq_r} + \frac{dT_{\dot{q}}}{dq_r} = 0.$$

* This demonstration is taken from Maxwell's 'Electricity.'

In the foregoing demonstration it is assumed that the time does not enter explicitly into the connecting equations, and therefore that T_p and $T_{\dot q}$ are homogeneous quadratic functions of the p's and $\dot q$'s respectively; in which case, as has just now been proved, the equation $\frac{dT_p}{dp_r} = \dot q_r$ follows as an analytical consequence from the proposition $\frac{dT_{\dot q}}{d\dot q_r} = p_r$. It may however be demonstrated independently, and *whether the time enters into the connecting equations or not*, that $\frac{dT_p}{dp_r} = \dot q_r$, it being borne in mind, as in the ~~last~~ first Proposition, that T_p ceases to be a homogeneous quadratic function of the p's in such a case.

For $$p_r = \Sigma m\left(\frac{dx}{dq_r}\frac{dx}{dt} + \frac{dy}{dq_r}\frac{dy}{dt} + \frac{dz}{dq_r}\frac{dz}{dt}\right).$$

Let a small impulse act on the system, whereby the velocity of each element is varied without change of position.

Then the quantities $\frac{dx}{dt}, \frac{dy}{dt}, \frac{dz}{dt}$ will vary, while $\frac{dx}{dq_r}, \frac{dy}{dq_r}, \frac{dz}{dq_r}$ remain constant;

$$\therefore\quad \delta p_r = \Sigma m\left(\frac{dx}{dq_r}\delta\frac{dx}{dt} + \frac{dy}{dq_r}\delta\frac{dy}{dt} + \frac{dz}{dq_r}\delta\frac{dz}{dt}\right);$$

$$\therefore\quad \dot q_r\,\delta p_r = \Sigma m\left(\dot q_r\frac{dx}{dq_r}\delta\frac{dx}{dt} + \dot q_r\frac{dy}{dq_r}\delta\frac{dy}{dt} + \dot q_r\frac{dz}{dq_r}\delta\frac{dz}{dt}\right);$$

$$\therefore\quad \Sigma\dot q\,\delta p$$

$$= \Sigma m\left\{\left(\dot q_1\frac{dx}{dq_1} + \dot q_2\frac{dx}{dq_2} + \&c.\right)\delta\frac{dx}{dt} + \left(\dot q_1\frac{dy}{dq_1} + \dot q_2\frac{dy}{dq_2} + \&c.\right)\delta\frac{dy}{dt}\right.$$
$$\left. + \left(\dot q_1\frac{dz}{dq_1} + \dot q_2\frac{dz}{dq_2} + \&c.\right)\delta\frac{dz}{dt}\right\}$$

$$= \Sigma m\left(\frac{dx}{dt}\delta\frac{dx}{dt} + \frac{dy}{dt}\delta\frac{dy}{dt} + \frac{dz}{dt}\delta\frac{dz}{dt}\right) = \delta T = \Sigma\frac{dT_p}{dp}\delta p.$$

And the δp's being independent,

$$\therefore\quad \frac{dT_p}{dp_r} = \dot q_r.$$

PROPOSITION IV. $\frac{dp_r}{d\dot{q}_s} = \frac{dp_s}{d\dot{q}_r}$.

Since $T_{\dot{q}}$ is a homogeneous quadratic function of the $\dot{q}$'s it must contain a term of the form $C\dot{q}_r\dot{q}_s$, where C is independent of the $\dot{q}$'s.

$$\therefore \quad p_r = \frac{dT_{\dot{q}}}{d\dot{q}_r} = C\dot{q}_s + \text{terms independent of } \dot{q}_s.$$

Similarly $$p_s = \frac{dT_{\dot{q}}}{d\dot{q}_s} = C\dot{q}_r + \text{terms independent of } \dot{q}_r;$$

$$\therefore \quad \frac{dp_r}{d\dot{q}_s} = C = \frac{dp_s}{d\dot{q}_r}.$$

PROPOSITION V. *If $p\dot{q}$ and $p'\dot{q}'$ represent two different states of motion of the system in the same configuration,*

$$\Sigma p\dot{q}' = \Sigma p'\dot{q}.$$

If variations δq_1, δq_2, &c. of the coordinates give rise to the displacement δr in any particle m whose velocity is v, we know from definition that

$$\Sigma p\delta q = \Sigma m v \delta r \cos a,$$

where a is the angle between v and δr.

When the system is in the p, $\dot{q}$ state, let the variations δq_1, δq_2, &c. be given to the coordinates such that

$$\delta q_1 = \dot{q}_1{}' \delta t; \quad \delta q_2 = \dot{q}_2{}' dt, \text{ \&c.}$$

Then it follows that $\delta r = v'\delta t$,

$$\therefore \quad \Sigma p\dot{q}'\delta t = \Sigma m v v' \cos a \delta t.$$

Similarly $$\Sigma p'\dot{q}\delta t = \Sigma m v' v \cos a \delta t;$$

$$\therefore \quad \Sigma p\dot{q}' = \Sigma p'\dot{q}.$$

This proposition may also be deduced from the linear equations connecting the p's and $\dot{q}$'s.

PROPOSITION VI. *If $p\dot{q}$, $p'\dot{q}'$ represent momenta and velocities in two distinct motions of a system in the same configuration, then if $\dot{q} + \dot{q}'$ represent velocities in a third state of motion with the same configuration, $p + p'$ will represent the momenta in this third state.*

This follows from the fact that the p's are homogeneous linear functions of the $\dot{q}$'s with coefficients known functions of the q's.

PROPOSITION VII. *If $\dot q$, $\dot q'$ and $\dot q + \dot q'$ represent component velocities of any system in the same configuration but different states of motion, and if the notation $T_{\dot q}$ represent kinetic energy corresponding to the state $\dot q$,*

$$T_{\dot q + \dot q'} = T_{\dot q} + T_{\dot q'} + \Sigma p \dot q',$$
$$= T_{\dot q} + T_{\dot q'} + \Sigma p' \dot q.$$

Since $T_{\dot q}$ is a homogeneous quadratic function of the n variables $\dot q_1$, $\dot q_2$ &c., it follows from Taylor's theorem applied to any number of variables that

$$T_{\dot q + \dot q'} = T_{\dot q} + \Sigma \frac{dT_{\dot q}}{d\dot q} \dot q' + R,$$

where R is independent of the $\dot q$'s and therefore by symmetry must be equal to $T_{\dot q}'$.

Also $$\frac{dT_{\dot q}}{d\dot q} = p;$$

$$\therefore \quad T_{\dot q + \dot q'} = T_{\dot q} + T_{\dot q'} + \Sigma p \dot q',$$
$$= T_{\dot q} + T_{\dot q'} + \Sigma p' \dot q.$$

Similarly $$T_{\dot q - \dot q'} = T_{\dot q} + T_{\dot q'} - \Sigma p \dot q',$$
$$= T_{\dot q} + T_{\dot q'} - \Sigma p' \dot q;$$

with similar propositions concerning

$$T_{p+p'} \text{ and } T_{p-p'}.$$

PROPOSITION VIII. *If $\Sigma p \dot q = \Sigma p' \dot q'$, then either is greater than $\Sigma p' \dot q$ or $\Sigma p \dot q'$.*

For if $\dot q - \dot q'$ represent the velocities in a third state of motion, $p - p'$ represent the momenta in that state by Prop. VI.

Therefore $\frac{1}{2}\Sigma (p - p')(\dot q - \dot q')$ represents the kinetic energy in that state and is therefore positive;

that is $\Sigma p \dot q + \Sigma p' \dot q' - \Sigma p' \dot q - \Sigma p \dot q'$ is positive;

that is $2\,\Sigma p \dot q - 2\,\Sigma p' \dot q$ is positive;

and therefore $\Sigma p \dot q$ or $\Sigma p' \dot q'$ is greater than $\Sigma p' \dot q$ or $\Sigma p \dot q'$.

It appears from this that $\Sigma p \dot q - \Sigma p \dot q'$ can never be zero unless $\dot q = \dot q'$ for each coordinate.

10.] If $f(q_1 \dots q_n)$, or shortly f, be any function of the coordinates, and $\frac{df}{dt}$ or $\dot{f}$ its rate of increase per unit of time, then, if $\dot{f}$ be given, T is the least possible, and if T be given, $\dot{f}^2$ is the greatest possible, when for each coordinate p is proportional to $\frac{df}{dq}$.

For let $\dot{q}$, p be a set of velocities and momenta in a motion such that $\Sigma \dot{q} \frac{df}{dq}$, or $\frac{df}{dt}$, has the given value, and that $p = \lambda \frac{df}{dq}$, where λ is some constant.

Let $\dot{q}+\dot{q}'$, $p+p'$ be the velocities and momenta in another motion which gives the same value for $\frac{df}{dt}$, and therefore such that

$$\Sigma(\dot{q}+\dot{q}')\frac{df}{dq} = \Sigma\dot{q}\frac{df}{dq},$$

and therefore

$$\Sigma\dot{q}'\frac{df}{dq} = 0.$$

Then for the doubled kinetic energy of this second motion we have

$$\begin{aligned} 2T &= \Sigma(p+p')(\dot{q}+\dot{q}'), \\ &= \Sigma p\dot{q} + \Sigma p'\dot{q}' + 2\Sigma p\dot{q}', \\ &= \Sigma p\dot{q} + \Sigma p'\dot{q}' + 2\lambda\Sigma\dot{q}'\frac{df}{dq}, \\ &= \Sigma p\dot{q} + \Sigma p'\dot{q}', \end{aligned}$$

which exceeds $\Sigma p\dot{q}$, the doubled kinetic energy of the p, $\dot{q}$ motion, by $\Sigma p'\dot{q}'$, an essentially positive quantity.

In order to find the actual value of this least kinetic energy, we must express every $\dot{q}$ in terms of the p's in the equation

$$\dot{f} = \Sigma\dot{q}\frac{df}{dq},$$

and then substitute

$$\lambda\frac{df}{dq} \text{ for } p.$$

If the linear equations expressing $\dot{q}$ in terms of the p's be of the form

$$\dot{q}_1 = B_{11}p_1 + B_{12}p_2 + \dots,$$

the result is

$$\dot{f} = \lambda \left\{ B_{11} \left(\frac{df}{dq_1}\right)^2 + 2 B_{12} \frac{df}{dq_1}\frac{df}{dq_2} + \dots \right\},$$

$$= \lambda F, \text{ suppose;}$$

whence $\lambda = \frac{\dot{f}}{F}$ and $2T = \frac{\dot{f}^2}{F}$.

F is the expression for $2T_p$ with $\frac{df}{dq}$ written for p.

Secondly let T be given, and let $\dot{q}, p$ be a set of velocities and momenta in a motion such that T has the given value, and that $p = \lambda \frac{df}{dq}$, and let $\dot{q} + \dot{q}', p + p'$ be those in any other motion having the same kinetic energy. Then

$$\Sigma \overline{p+p'}\ \overline{\dot{q}+\dot{q}'} - \Sigma p\dot{q} = 0,$$

or

$$\Sigma p'\dot{q}' + 2\Sigma p\dot{q}' = 0. \quad \dots \dots \dots \quad (1)$$

Therefore

$$\left(\Sigma \frac{df}{dq}\dot{q}\right)^2 - \left(\Sigma \frac{df}{dq}(\dot{q}+\dot{q}')\right)^2 = \frac{1}{\lambda^2}\left\{ (\Sigma p\dot{q})^2 - \left(\Sigma p(\dot{q}+\dot{q}')\right)^2 \right\},$$

$$= -\frac{1}{\lambda^2}\{(\Sigma p\dot{q}')^2 + 2\Sigma p\dot{q}\Sigma p\dot{q}'\},$$

$$= -\frac{1}{\lambda^2}\Sigma p\dot{q}'\{\Sigma p\dot{q}' + 2\Sigma p\dot{q}\}$$

$$= +\frac{1}{\lambda^2}\frac{\Sigma p'\dot{q}'}{2}\{\Sigma p\dot{q}' + 2\Sigma p\dot{q}\}. \quad \dots \dots \quad (2)$$

Now let $\dot{q}''$ denote a set of velocities proportional to the $\dot{q}'$ set so that $\dot{q}' = r\dot{q}''$, where r is some numerical quantity, and consequently $p' = rp''$. Let further r be so chosen that

$$\Sigma p''\dot{q}'' = \Sigma p\dot{q}.$$

Then from (1)

$$r = -\frac{2\Sigma p\dot{q}''}{\Sigma p\dot{q}}.$$

And (2) becomes

$$\left(\Sigma \frac{df}{dq}\dot{q}\right)^2 - \left(\Sigma \frac{df}{dq}(\dot{q}+\dot{q}')\right)^2 = \frac{1}{\lambda^2}\frac{\Sigma p'\dot{q}'}{2}(r\Sigma p\dot{q}'' + 2\Sigma p\dot{q})$$

$$= \frac{1}{\lambda^2}\Sigma p'\dot{q}'\frac{(\Sigma p\dot{q})^2 - (\Sigma p\dot{q}'')^2}{\Sigma p\dot{q}},$$

which is necessarily positive by Proposition VIII. Therefore

$$(\Sigma \frac{df}{dq}\dot{q})^2 \text{ or } \dot{f}^2 \text{ is greater than } \left(\Sigma \frac{df}{dq}(\dot{q}+\dot{q}')\right)^2.$$

11.] If p_r and F_r be generalised components of momentum and force corresponding to any coordinate q_r, then

$$\frac{dp_r}{dt} - \frac{dT_{\dot{q}}}{dq_r} = F_r.$$

We have seen in Art. 2 that

$$p_r = \Sigma m\left(\frac{dx}{dt}\cdot\frac{dx}{dq_r} + \frac{dy}{dt}\cdot\frac{dy}{dq_r} + \frac{dz}{dt}\cdot\frac{dz}{dq_r}\right), \quad \ldots \ldots \ldots (1)$$

and that
$$F_r = \Sigma m\left(\frac{d^2x}{dt^2}\cdot\frac{dx}{dq_r} + \frac{d^2y}{dt^2}\cdot\frac{dy}{dq_r} + \frac{d^2z}{dt^2}\cdot\frac{dz}{dq_r}\right). \quad \ldots \ldots \ldots (2)$$

Therefore by differentiation of (1), remembering (2),

$$\frac{dp_r}{dt} = F_r + \dot{\Sigma} m\left(\frac{dx}{dt}\cdot\frac{d}{dt}\cdot\frac{dx}{dq_r} + \frac{dy}{dt}\cdot\frac{d}{dt}\cdot\frac{dy}{dq_r} + \frac{dz}{dt}\cdot\frac{d}{dt}\cdot\frac{dz}{dq_r}\right).$$

Now

$$\frac{d}{dt}\cdot\frac{dx}{dq_r} = \dot{q}_1\frac{d^2x}{dq_1 dq_r} + \dot{q}_2\frac{d^2x}{dq_2 dq_r} + \&c. = \frac{d}{dq_r}\cdot\left(\dot{q}_1\frac{dx}{dq_1} + \dot{q}_2\frac{dx}{dq_2} + \&c.\right)$$
$$= \frac{d}{dq_r}\cdot\frac{dx}{dt},$$

and similarly for $\frac{d}{dt}\frac{dy}{dq_r}$ and $\frac{d}{dt}\frac{dz}{dq_r}$;

$$\therefore \quad \frac{dp_r}{dt} = F_r + \tfrac{1}{2}\cdot\frac{d}{dq_r}\cdot\Sigma m\left\{\left(\frac{dx}{dt}\right)^2 + \left(\frac{dy}{dt}\right)^2 + \left(\frac{dz}{dt}\right)^2\right\},$$

provided that in differentiating with regard to q_r the $\dot{q}$'s remain constant;

$$\therefore \quad \frac{dp_r}{dt} = F_r + \frac{dT_{\dot{q}}}{dq_r},$$

or
$$\frac{dp_r}{dt} - \frac{dT_{\dot{q}}}{dq_r} = F_r.$$

Since, by Proposition I, $\frac{dT_{\dot{q}}}{d\dot{q}_r} = p_r$, this result may be written in the form

$$\frac{d}{dt}\cdot\left(\frac{dT_{\dot{q}}}{d\dot{q}_r}\right) - \frac{dT_{\dot{q}}}{dq_r} = F_r.$$

And since the demonstrations of this proposition as well as that of Prop. I hold when the time enters explicitly into the connecting equations, *we have in all cases*

$$\frac{dp_r}{dt} - \frac{dT_{\dot q}}{dq_r} = \frac{d}{dt}\cdot\left(\frac{dT_{\dot q}}{d\dot q_r}\right) - \frac{dT_{\dot q}}{dq} = F_r,$$

remembering that $T_{\dot q}$ ceases to be a homogeneous quadratic function of the $\dot q$'s when the time thus enters.

By Proposition III it follows that when the time does not enter explicitly the result may be written

$$\frac{dp_r}{dt} - \frac{dT_{\dot q}}{dq_r} = \frac{dp_r}{dt} + \frac{dT_p}{dq_r} = F_r.$$

12.] These are called Lagrange's Equations of Motion. They are applicable to systems moving under the influence of finite forces only. The corresponding forms for impulsive forces are easily deduced from the foregoing propositions. For if P be the generalised component of impulses acting on the system corresponding to the coordinate q, we have seen that if the motion be from rest $P = p$.

But the velocities created in the system by any impulses are irrespective of the state of the system as regards rest or motion at the time when the impulses act. Therefore if p_0 denote the momenta before, and p after, the impulses, $\dot q_0$ and $\dot q$ the corresponding velocity components, we must have

$$P = p - p_0,$$

or

$$\frac{dT}{d\dot q} - \frac{dT}{d\dot q_0} = P.$$

13.] The following are examples of the use of Lagrange's Equations.

Example 1. The system of two pulleys in Example 1, Art. 8 moving from rest under the action of gravity; it is required to determine the motion.

Evidently in this case

$$\frac{dT}{dq_1} = 0, \quad \frac{dT}{dq_2} = 0;$$

therefore

$$\frac{dp_1}{dt} = -\frac{dU}{dq_1} = (m_1 - m_2 - m_3)\,g,$$

$$\frac{dp_2}{dt} = -\frac{dU}{dq_2} = (m_2 - m_3)\,g;$$

where V is the potential energy.

therefore $$p_1 = (m_1 - m_2 - m_3)\, gt,$$
$$p_2 = (m_2 - m_3)\, gt;$$
no constants of integration being required since the motion is from rest.

Hence, substituting for p_1, p_2 their values given in Example 1, Art. 8, we obtain

$$(m_1+m_2+m_3)\,\dot{q}_1 + (m_3 - m_2)\,\dot{q}_2 = (m_1 - m_2 - m_3)\, gt,$$
$$(m_3 - m_2)\,\dot{q}_1 + (m_2 + m_3)\,\dot{q}_2 = (m_2 - m_3)\, gt;$$

that is,

$$(m_1+m_2+m_3)(q_1 - q_1') + (m_3 - m_2)(q_2 - q_2') = (m_1 - m_2 - m_3)\frac{gt^2}{2},$$
$$(m_3 - m_2)(q_1 - q_1') + (m_2 + m_3)(q_2 - q_2') = (m_2 - m_3)\frac{gt^2}{2};$$

q_1', q_2' being the initial lengths of the strings measured from the vertices of the pulleys.

Hence $$q_1 - q_1' = \frac{m_1(m_2+m_3) - 4\,m_2 m_3}{m_1(m_2+m_3) + 4\,m_2 m_3}\,\frac{gt^2}{2},$$

$$q_2 - q_2' = \frac{2\,m_1(m_2 - m_3)}{m_1(m_2+m_3) + 4\,m_2 m_3}\,\frac{gt^2}{2};$$

and $\dot{q}_1$ is found from $q_1 - q_1'$ by substituting gt for $\frac{gt^2}{2}$.

In the same manner let there be $\lambda - 1$ moveable pulleys, and let the $\lambda + 1$ weights be all equal; we shall obtain

$$(\lambda+1)\,\dot{q}_1 + (\lambda - 2)\,\dot{q}_2 + (\lambda - 3)\,\dot{q}_3 + \ldots + \dot{q}_{\lambda-1} = p_1 = -(\lambda - 1)\, gt,$$
$$(\lambda - 2)\,\dot{q}_1 + \lambda\,\dot{q}_2 + (\lambda - 3)\,\dot{q}_3 + \ldots + \dot{q}_{\lambda-1} = p_2 = -(\lambda - 2)\, gt.$$
$$\&c. = \&c.$$

From which any $\dot{q}$ can be expressed in the form of a determinant.

Example 2. The following is taken from Routh's Rigid Dynamics.

To deduce Euler's equations for a rigid body from Lagrange's equations of motion.

We have shewn above that if p_θ, p_ϕ, p_ψ denote the generalised components of momentum corresponding to the coordinates θ, ϕ, ψ respectively,

$$p_\theta = B\omega_2 \cos\phi + A\,\omega_1 \sin\phi,$$
$$p_\phi = + C\omega_3,$$
$$p_\psi = C\omega_3 \cos\theta + (B\omega_2 \sin\phi - A\,\omega_1 \cos\phi)\sin\theta.$$

But by Lagrange's equations

$$\frac{dp_\phi}{dt} - \frac{dT}{d\phi} = -\frac{dU}{d\phi};$$

that is

$$+C\frac{d\omega_3}{dt} + (B\omega_2 \sin\phi - A\,\omega_1 \cos\phi)\,\dot{\theta},$$

$$-(B\omega_2 \cos\phi + A\,\omega_1 \sin\phi)\sin\theta\,\dot{\psi} = -\frac{dU}{d\phi}.$$

But

$$\dot{\theta} = \omega_1 \sin\phi + \omega_2 \cos\phi,$$

$$\sin\theta\,\dot{\psi} = -\omega_1 \cos\phi + \omega_2 \sin\phi.$$

Hence we obtain by substitution,

$$-\frac{dU}{d\phi} = +C\frac{d\omega_3}{dt} + (B\,\omega_2 \sin\phi - A\,\omega_1 \cos\phi)(\omega_1 \sin\phi + \omega_2 \cos\phi)$$

$$-(B\,\omega_2 \cos\phi + A\,\omega_1 \sin\phi)(-\omega_1 \cos\phi + \omega_2 \sin\phi)$$

$$= +C\frac{d\omega_3}{dt} + (B - A)\,\omega_1\,\omega_2,$$

and the two other equations of Euler's system are deducible from this by symmetry.

Example 3. A rigid body is supported on a fixed axis and another rigid body is supported on the first by another axis.

Case (*a*). If the second axis be parallel to the first. (Thomson and Tait's *Natural Philosophy*, § 330, p. 257.)

Here there are two degrees of freedom, and the coordinates may be conveniently taken to be (1) ϕ the inclination of the plane containing the axes to a fixed plane through the first axis, and (2) ψ the inclination of the fixed plane ~~of the axes~~ to a plane through the second axis and the centre of gravity of the second body.

With this notation it is easily seen that if a be the distance between the axes, and b the distance of the C. G. of the second body from the second axis, then the (velocity)^2 of the C. G. of the second body is

$$a^2\dot{\phi}^2 + 2ab\,\dot{\psi}\dot{\phi}\cos(\psi - \phi) + b^2\dot{\psi}^2;$$

and therefore, if m and m' be the masses of the bodies, and j and k their respective radii of gyration round the first axis and round an axis through the C. G. of the second body parallel to the two axes, then the kinetic energy T is such that

$$T = \tfrac{1}{2}\left\{mj^2\dot{\phi}^2 + m'\left[a^2\dot{\phi}^2 + 2ab\dot{\psi}\dot{\phi}\cos(\psi - \phi) + b^2\dot{\psi}^2 + k^2\dot{\psi}^2\right]\right\};$$

whence Lagrange's equations become

$$(mj^2+m'a^2)\ddot{\phi}+m'ab\,\frac{d}{dt}[\dot{\psi}\cos(\psi-\phi)]-m'ab\sin(\psi-\phi)\,\dot{\phi}\dot{\psi}=\Phi,$$

or $$(mj^2+m'a^2)\ddot{\phi}+m'ab\,\ddot{\psi}\cos\overline{\psi-\phi}-m'ab\,\dot{\psi}^2\sin\overline{\psi-\phi}=\Phi,$$

$$m'ab\frac{d}{dt}[\dot{\phi}\cos(\psi-\phi)]+m'(b^2+k^2)\,\ddot{\psi}+m'ab\sin(\psi-\phi)\,\dot{\phi}\dot{\psi}=\Psi;$$

where Φ and Ψ are the generalised force components corresponding to ϕ and ψ.

In case of gravity or other external force the potential energy may be readily obtained in terms of ϕ, ψ, and then Φ, Ψ may be found by differentiation.

Case (*b*). The axes at right angles to each other.

Let ϕ be the angle between the plane containing one of the axes and the shortest distance between them and some fixed plane containing the aforesaid axis.

Let ψ be the angle between the plane containing the second axis and the C. G. of the second body, and the plane containing the second axis and the shortest distance.

Let a be the shortest distance between the axes, r the length of the perpendicular let fall from the C. G. of the second body upon the second axis, and b the distance measured along this second axis between the foot of a and the foot of r.

Then it is easily seen that the velocities of the C. G. of the second body parallel to the first axis, to the second axis, and the shortest distance respectively are

$$r\cos\psi\,.\,\dot{\psi},\quad a\dot{\phi}+r\cos\psi\dot{\phi},\quad\text{and}\quad b\dot{\phi}+r\sin\psi\,.\,\dot{\psi}.$$

And therefore if m' be the mass of the second body the kinetic energy of that body arising from the motion of translation of its C. G. will be

$$\frac{m'}{2}\{(a+r\cos\psi)^2\,\dot{\phi}^2+(b\,\dot{\phi}+r\sin\psi\,.\,\dot{\psi})^2+r^2\cos^2\psi\,.\,\dot{\psi}^2\},$$

or $$\frac{m'}{2}\,.\,\{[(a+r\cos\psi)^2+b^2]\,\dot{\phi}^2+2\,br\sin\psi\,\dot{\psi}\dot{\phi}+r^2\,\dot{\psi}^2\}.$$

The kinetic energy of rotation of the second body is

$$A\,\omega_1^{\,2}+B\,\omega_2^{\,2}+C\,\omega_3^{\,2},$$

where A, B, C are its principal moments of inertia about the C. G., and ω_1, ω_2, ω_3 the rotations about the principal axes at that point.

The quantities ω_1, ω_2, ω_3 will be linear functions of $\dot\phi$ and $\dot\psi$ with coefficients functions of ψ, which can only be expressed when the circumstances of each particular case are known.

The kinetic energy of the first body is $\frac{1}{2} I\dot\phi^2$, where I is the moment of inertia of that body about the first axis.

As a particular case, suppose the C. G. of the second body to be situated in the plane containing the first axis and the shortest distance, i.e. suppose that $b = 0$.

Suppose also that the second axis is parallel to a principal axis through the C. G. of the second body.

Then the kinetic energy of translation of the second body becomes

$$\frac{m'}{2}\{(a+r\cos\psi)^2\dot\phi^2+r^2\dot\psi^2\}.$$

And the kinetic energy of rotation of that body becomes

$$\tfrac{1}{2}\left\{\{A\cos^2(\psi+a)+B\sin^2(\psi+a)\}\dot\phi^2+C\dot\psi^2\right\};$$

that of the first body being as before $\frac{1}{2} I\dot\phi^2$.

Therefore twice the kinetic energy of the whole system assumes the form

$$(P+Q\cos\psi+R\cos^2\psi+S\cos^2\overline{\psi+a})\,\dot\phi^2+U\dot\psi^2,$$

where P, Q, R, S, and U are known functions of the given constants.

And Lagrange's equations become

$$\frac{d}{dt}\cdot(\{P+Q\cos\psi+R\cos^2\psi+S\cos^2(\psi+a)\}\,\dot\phi) = \Phi,$$

$$U\ddot\psi-\dot\phi^2\frac{d}{d\psi}\{P+Q\cos\psi+R\cos^2\psi+S\cos^2(\psi+a)\} = \Psi;$$

Φ and Ψ being generalised components of force corresponding to ϕ and ψ respectively.

If the first axis be vertical, and if gravity be the only impressed force,

$$\Phi = 0 \text{ and } \Psi = -m'gr\cos\psi,$$

and therefore the equations become

$$\{P+Q\cos\psi+R\cos^2\psi+S\cos^2(\psi+a)\}\,\dot\phi = \text{E (const.)},$$

$$\ddot\psi-\frac{\mathrm{E}^2\dfrac{d}{d\psi}(P+Q\cos\psi+R\cos^2\psi+S\cos^2(\psi+a))}{(P+Q\cos\psi+R\cos^2\psi+S\cos^2(\psi+a))^2} = -m'gr\cos\psi;$$

or
$$\dot\psi^2+\frac{2\mathrm{E}^2}{(P+Q\cos\psi+R\cos^2\psi+S\cos^2(\psi+a))} = \frac{2m'gr(\sin\beta-\sin\psi)}{U},$$

giving $$\psi \text{ or } \frac{d\psi}{dt} = f(\psi);$$

whence t may be found as a function of ψ by mere integration, and therefore conversely ψ may be found as a function of t.

And then by substitution $\dot{\phi}$, and therefore ϕ, is found from the equation

$$\dot{\phi} = \frac{d\phi}{dt} = \frac{E}{P + Q\cos\psi + R\cos^2\psi + S\cos^2(\psi + a)}.$$

If the circumstances of the motion make $a = 0$, the expressions are slightly simplified by the two constants R and S blending into one.

14.] If F_r denote the generalised component of impressed force corresponding to the coordinate q_r, the work done per unit of time by the impressed forces on the system moving with the velocities $\dot{q}_1, \ldots \dot{q}_n$, that is the increase per unit of time of the kinetic energy, is $\Sigma F\dot{q}$; or by Lagrange's equations,

$$\Sigma \left(\frac{dp}{dt} - \frac{dT_{\dot{q}}}{dq}\right)\dot{q}.$$

Now $$\frac{dp_r}{dt} = \Sigma\dot{q}\frac{d}{dt}\frac{dp_r}{d\dot{q}} + \Sigma\frac{dp_r}{d\dot{q}}\ddot{q};$$

$$\begin{aligned} \therefore \quad \Sigma\dot{q}\frac{dp}{dt} &= \dot{q}_1\left\{\Sigma\dot{q}\frac{d}{dt}\frac{dp_1}{d\dot{q}} + \Sigma\frac{dp_1}{d\dot{q}}\ddot{q}\right\} \\ &+ \dot{q}_2\left\{\Sigma\dot{q}\frac{d}{dt}\frac{dp_2}{d\dot{q}} + \Sigma\frac{dp_2}{d\dot{q}}\ddot{q}\right\} \\ &+ \quad \cdot \quad \cdot \quad \cdot \quad \cdot \end{aligned}$$

and $$\Sigma\dot{q}\frac{dT_{\dot{q}}}{dq} = \tfrac{1}{2}\dot{q}_1\Sigma\dot{q}\frac{d}{dt}\frac{dp_1}{d\dot{q}} + \tfrac{1}{2}\dot{q}_2\Sigma\dot{q}\frac{d}{dt}\frac{dp_2}{d\dot{q}} + \ldots.$$

Therefore $$\begin{aligned} \Sigma F\dot{q} &= \Sigma\dot{q}\left(\frac{dp}{dt} - \frac{dT_{\dot{q}}}{dq}\right) \\ &= \dot{q}_1\Sigma\frac{dp_1}{d\dot{q}}\ddot{q} + \dot{q}_2\Sigma\frac{dp_2}{d\dot{q}}\ddot{q} + \ldots \\ &+ \Sigma\dot{q}\frac{dT_{\dot{q}}}{dq}. \end{aligned}$$

If the velocities be indefinitely small the last term may be neglected in comparison with the others, because it involves only higher powers of the $\dot{q}$'s. In that case, but generally not otherwise, we may equate the coefficients of each $\dot q$, and obtain for a system at rest,

$$F_r = \Sigma \frac{dp_r}{d\dot q}\ddot q.$$

If the velocities $\dot q_1, \dots \dot q_r$ only be reduced to zero, the work done per unit of time on the system moving with the remaining velocities $\dot q_{r+1} \dots \dot q_n$ becomes by arrangement of the terms

$$\Sigma^n_{r+1} F\dot q = \dot q_{r+1}\Sigma_1^r \frac{dp_{r+1}}{d\dot q}\ddot q + \dot q_{r+2}\Sigma_1^r \frac{dp_{r+2}}{d\dot q}\ddot q + \dots$$

$$+ \dot q_{r+1}\Sigma^n_{r+1}\frac{dp_{r+1}}{d\dot q}\ddot q + \dot q_{r+2}\Sigma^n_{r+1}\frac{dp_{r+2}}{d\dot q}\ddot q + \dots + \Sigma^n_{r+1}\dot q\frac{dT_{\dot q}}{dq};$$

or $$\dot q_{r+1}\left\{F_{r+1} - \Sigma_1^r\frac{dp_{r+1}}{d\dot q}\ddot q\right\} + \dot q_{r+2}\left\{F_{r+2} - \Sigma_1^r\frac{dp_{r+2}}{d\dot q}\ddot q\right\} + \&c.,$$

$$= \dot q_{r+1}\Sigma^n_{r+1}\frac{dp_{r+1}}{d\dot q}\ddot q + \dot q_{r+2}\Sigma^n_{r+1}\frac{dp_{r+2}}{d\dot q}\ddot q + \&c.,$$

$$+ \Sigma^n_{r+1}\dot q\frac{dT_{\dot q}}{dq}.$$

Now let $q_1 \dots q_r$ define the position of a moving space, $q_{r+1} \dots q_n$ that of a system moving relatively to the space. In that case the second member of the last equation expresses the increase per unit of time of the kinetic energy of the relative motion. And the equation shews that this is obtained by subtracting from any component of force—e. g. F_{r+1}—the quantity $\Sigma^r_1 \frac{dp_{r+1}}{d\dot q}\ddot q$, which is what the generalised component of effective force corresponding to q_{r+1} would be if the space were at rest and the system fixed to it, $\ddot q_{r+1} \dots \ddot q_n$ being therefore zero. Hence we might deduce ~~This expresses~~ Coriolis' theorem.

15.] To find the work done by any impulse acting on a material system in any given state of motion.

Let $\dot q$, p represent any components of velocity and momentum before the impulse acts, and let $\dot q + \dot q'$ and $p + p'$ be the corresponding components after the impulse.

Let P' be any component of the impulse corresponding to the above-mentioned components of velocity and momentum.

Then employing the notation of the preceding articles we know that the work done by the impulse must be equal to

$$T_{p+p',\ q+q'} - T_{pq},$$

or

$$T_p + T_{p'} + \Sigma p'\dot{q} - T_p,$$

or

$$\tfrac{1}{2}\Sigma p'\dot{q}' + \Sigma p'\dot{q}.$$

Let the new velocity $\dot{q}+\dot{q}'$ be denoted by $\dot{Q}$, then this expression for the work becomes

$$\Sigma p'\dot{q} + \Sigma\tfrac{1}{2}p'(\dot{Q}-\dot{q}),$$

$$= \Sigma p'\frac{\dot{q}+\dot{Q}}{2}.$$

Also by D'Alembert's principle $p' = P'$.

Therefore the work done by the impulse whose generalised components are P_1, P_2, &c., is

$$\Sigma P\frac{\dot{q}+\dot{Q}}{2}.$$

If the impulse whose rectangular components are X, Y, Z act at a point of the system whose coordinates are x, y, z and component velocities u, v, w, and if U, V, W be the values of u, v, w after the impulse, then the work done will be found by substituting X, Y, Z for P_1, P_2, P_3 in the above expression and making each of the remaining components, P_4, P_5, &c., zero; so that it becomes

$$X\frac{u+U}{2} + Y\frac{v+V}{2} + Z\frac{w+W}{2}.$$

CHAPTER II.

MAXIMUM AND MINIMUM KINETIC ENERGY.

ARTICLE 16.] *If any system at rest in any configuration be acted on by any given impulses, the kinetic energy imparted will be greater the greater the number of degrees of freedom of the system. And for every additional constraint introduced there will be a loss of kinetic energy equal to that of the motion which, compounded with the unconstrained, would produce the constrained motion.* (Bertrand's Theorem.)

For let P_1, P_2, &c. be the generalised components of impulse acting on the system. Let $\dot{q}_1$, $\dot{q}_2$, &c. be the resulting components of velocity, p_1, p_2, &c. the corresponding momenta, and T the kinetic energy. Then by what has been already proved, we know that

$$P_1 = p_1, \quad P_2 = p_2, \text{ \&c., and } \quad T = \tfrac{1}{2}\Sigma p\dot{q}.$$

Let any constraint, which we may denote by C, be introduced into the system, such that when the same impressed impulses act upon it as before, the velocities and momenta in the constrained motion shall be $\dot{q}_1'$, $\dot{q}_2'$, &c. and p_1', p_2', &c., and the kinetic energy $T' = \frac{1}{2}\Sigma p'\dot{q}'$.

In the constrained system the possible displacements ∂q_1, ∂q_2, &c. are no longer independent, but it is still true by D'Alembert's principle, Art. 6, *note*, that if ∂q_1, ∂q_2, &c. represent any possible values of these displacements in the constrained system,

$$\Sigma(P - p')\partial q = 0,$$

although we cannot, as in the former case when the ∂q's were independent, equate the coefficient of each ∂q to zero, and deduce the equations $P_1 = p'_1$, $P_2 = p'_2$, &c. It is clear that if we take δq_1, δq_2, &c. proportional to $\dot{q}_1'$, $\dot{q}_2'$, &c., such values will be consistent values of the δq's, and therefore

$$\Sigma(P-p')\dot{q}'=0,$$

or $$\Sigma p'\dot{q}'=\Sigma P\dot{q}'=\Sigma p\dot{q}'=\Sigma p'\dot{q}.$$

Therefore $$T'=\tfrac{1}{2}\Sigma p'\dot{q}'=\tfrac{1}{2}\Sigma p\dot{q}'.$$

And
$$\begin{aligned} T-T' &= \tfrac{1}{2}\{\Sigma p\dot{q}-\Sigma p\dot{q}'\} \\ &= \tfrac{1}{2}\Sigma p(\dot{q}-\dot{q}') \\ &= \tfrac{1}{2}\Sigma(p-p')(\dot{q}-\dot{q}'), \end{aligned}$$

since $$\Sigma p'(\dot{q}-\dot{q}')=\Sigma p'\dot{q}-\Sigma p'\dot{q}'=0.$$

That is, $$T-T'=T_{\dot{q}-\dot{q}'}.$$

The motion which has to be combined with the free in order to produce the constrained motion, that is the motion $\dot{q}'-\dot{q}$, may be called *the constraining motion.*

17.] It follows as a corollary that the kinetic energy of the constraining motion $\dot{q}'-\dot{q}$ is less than that of any other motion which, compounded with the free motion, would cause the system to obey the constraint C: in other words, $T_{\dot{q}-\dot{q}'}$, the kinetic energy lost by the introduction of the constraint C, is the least possible. This is Gauss' principle of Least Constraint.

Let $\dot{q}'$ denote the velocities which the system when subjected to the constraint C, and to no other constraint, actually takes under the given impulses.

Let $\dot{q}''$ denote the velocities in any motion whatever which the system can have consistently with the constraint C.

Then as we have seen, by D'Alembert's principle

$$\left.\begin{aligned} \Sigma(P-p')\dot{q}''=\Sigma(p-p')\dot{q}''&=0, \\ \text{and therefore also} \qquad \Sigma(\dot{q}-\dot{q}')p''&=0. \end{aligned}\right\} \quad \ldots\ldots\ldots \quad (1)$$

Then $\dot{q}'+\dot{q}''-\dot{q}$ represents the velocities in any motion, different from $\dot{q}'-\dot{q}$, which, when compounded with the free motion, satisfies the constraint C. And we have

$$\begin{aligned} T_{\dot{q}'+\dot{q}''-\dot{q}}-T_{\dot{q}'-\dot{q}} &= \tfrac{1}{2}\Sigma(p'+p''-p)(\dot{q}'+\dot{q}''-\dot{q})-\tfrac{1}{2}\Sigma(p'-p)(\dot{q}'-\dot{q}) \\ &= \tfrac{1}{2}\Sigma(p'-p)\dot{q}''+\tfrac{1}{2}\Sigma(\dot{q}'-\dot{q})p''+\tfrac{1}{2}\Sigma p''\dot{q}'' \\ &= \tfrac{1}{2}\Sigma p''\dot{q}'' \text{ by (1)}; \end{aligned}$$

and this is necessarily positive, therefore

$$T_{\dot{q}'-\dot{q}}<T_{\dot{q}'+\dot{q}''-\dot{q}}.$$

This proposition sometimes admits of practical application if it be required to find the constrained motion when the free motion is known. If, for instance, only one degree of freedom be removed by the constraint, then the constraint may be expressed by making some one function of the coordinates constant in the constrained, which is not constant in the free, motion. If f be that function, $\Sigma \frac{df}{dq}\dot{q}$, the rate of increase of f per unit of time in the free motion, is known. In the constrained motion f is to be constant, that is

$$\Sigma \frac{df}{dq}\dot{q}' = 0;$$

whence

$$\Sigma \frac{df}{dq}(\dot{q}' - \dot{q}) = -\Sigma \frac{df}{dq}\dot{q}.$$

Now the kinetic energy of the $\dot{q}' - \dot{q}$ motion is, as we have just seen, less than that of any other motion which, combined with the free motion, satisfies the constraint; that is, in which the rate of increase of f per unit of time is

$$-\Sigma \frac{df}{dq}\dot{q}.$$

Therefore by Art. 10,

$$p' - p = \lambda \frac{df}{dq},$$

from which $p' - p$ may be determined as in Art. 10.

For example, two free particles of masses, m_1, m_2, move from rest under given impulses with velocities $\dot{x}_1, \dot{y}_1, \dot{z}_1, \dot{x}_2, \dot{y}_2, \dot{z}_2$. It is required to determine the velocities with which they will move off under the same impulses if constrained to remain at a constant distance, r, apart by being connected by a string or rod without mass. If $\dot{x}_1'$, &c. be the new velocities, we have

$$m_1 \{\dot{x}_1' - \dot{x}_1\} = \lambda \frac{dr}{dx_1}, \text{ \&c.}$$

And determining λ as in Art. 10, we find

$$\lambda = \frac{-\frac{dr}{dt}}{\frac{1}{m_1}\left\{\left(\frac{dr}{dx_1}\right)^2 + \left(\frac{dr}{dy_1}\right)^2 + \left(\frac{dr}{dz_1}\right)^2\right\} + \frac{1}{m_2}\left\{\left(\frac{dr}{dx_2}\right)^2 + \left(\frac{dr}{dy_2}\right)^2 + \left(\frac{dr}{dz_2}\right)^2\right\}},$$

where $\frac{dr}{dt}$ is the rate of increase of r with the time in the free motion; that is,

$$\lambda = \frac{m_1 m_2}{m_1 + m_2} \frac{dr}{dt},$$

since $$\left(\frac{dr}{dx_1}\right)^2 + \left(\frac{dr}{dy_1}\right)^2 + \left(\frac{dr}{dz_1}\right)^2 = \frac{(x_1 - x_2)^2}{r^2} + \frac{(y_1 - y_2)^2}{r^2} + \frac{(z_1 - z_2)^2}{r^2}$$

$$= 1,$$

and similarly $$\left(\frac{dr}{dx_2}\right)^2 + \left(\frac{dr}{dy_2}\right)^2 + \left(\frac{dr}{dz_2}\right)^2 = 1.$$

Hence $$\dot{x}_1' - \dot{x}_1 = \frac{m_2}{m_1 + m_2} \frac{x_1 - x_2}{r} \frac{dr}{dt}$$

$$\dot{x}_2' - \dot{x}_2 = -\frac{m_1}{m_1 + m_2} \frac{x_1 - x_2}{r} \frac{dr}{dt}$$

&c. = &c.

The general problem of determining the constrained motion, when the free motion and the nature of the constraint are known, is more conveniently treated under the principle of least kinetic energy hereafter discussed. For every constraint must act at some definite point or points of the system, and may be conceived to consist in giving to these points certain new velocities in addition to the velocities which they take in the free motion. The kinetic energy of the constraining motion is then, as will be proved presently, the least which the system can have consistently with those points having the required new velocities. And this property, as will appear, suffices to determine the whole motion.

18.] The proposition proved in Art. 16 has been put into a somewhat more general form by Lord Rayleigh in the *Phil. Mag.*, vol. xlix. § 4, which, expressed in the language of generalised coordinates, is as follows.

Let P_1, P_2, &c. be any generalised components of impulse acting on any material system.

Let $\dot{Q}_1$, $\dot{Q}_2$, &c. be any possible quantities whatever, and let $T_{\dot{Q}}$ be the value of the kinetic energy of the system, when with the given configuration the velocity components are $\dot{Q}_1$, $\dot{Q}_2$, &c.

Let the expression $\Sigma P\dot{Q} - T_{\dot{Q}}$ be denoted by the symbol Ψ, and let ψ be the value of Ψ when for $\dot{Q}_1$, $\dot{Q}_2$, &c. have been substituted the values $\dot{q}_1$, $\dot{q}_2$, &c. of the component velocities actually assumed by the system at rest in the given configuration when acted on by the given impulses.

Then ψ is the greatest possible value of Ψ. For if p_1, p_2, &c. be the momenta actually assumed, we know that

$$P_1 = p_1, \quad P_2 = p_2, \quad \text{\&c.};$$

$$\therefore \quad \psi = \Sigma p\dot{q} - T_{\dot{q}} = 2T_{\dot{q}} - T_{\dot{q}} = T_{\dot{q}};$$

$$\therefore \quad \psi - \Psi = T_{\dot{q}} + T_{\dot{Q}} - \Sigma P\dot{Q}$$

$$= T_{\dot{q}} + T_{\dot{Q}} - \Sigma p\dot{Q}$$

$$= T_{\dot{q}-\dot{Q}}$$

by Proposition VII, and is therefore essentially positive.

The result of Art. 16 is a particular case of this proposition. For if $\dot{Q}_1$, $\dot{Q}_2$, &c. be the velocities assumed by the system when subjected to any constraint and acted on by the same impulses, Ψ is the kinetic energy assumed by the system, that is $T_{\dot{Q}}$, and the result just obtained assumes the form

$$T_{\dot{q}} = T_{\dot{Q}} + T_{\dot{q}-\dot{Q}},$$

the same as that of Art. 16.

19.] By the aid of the foregoing we may prove that when the masses of any part or parts of a material system are diminished, the connections and configuration being unaltered, the resulting kinetic energy under given impressed impulses from rest must be increased.

Substitute for the $\dot{Q}$'s in forming the function Ψ for the new system the values $\dot{q}_1$, $\dot{q}_2$, &c. of the velocities assumed under the given impulses in the old system, and let Ψ' be the value of Ψ thus found in the new system. Also let $T'_{\dot{q}}$ denote the kinetic energy of the new system corresponding to the velocities $\dot{q}_1 \dots \dot{q}_n$. Then

$$\Psi' = \Sigma(P\dot{q}) - T'_{\dot{q}},$$

where $\Sigma(P\dot{q})$ is the same as $\Sigma(P\dot{q})$ in the old system, but $T'_{\dot{q}}$ is clearly less than $T_{\dot{q}}$, because the configuration and velocities being the same the masses are diminished.

Therefore Ψ' is clearly greater than $\Sigma P\dot{q} - T_{\dot{q}}$ or $T_{\dot{q}}$. Let

ψ'' be the value of Ψ for the new system when the velocities actually assumed by it (supposed to be $\dot{q}'_1, \dot{q}'_2$, &c.) have been substituted for the $\dot{Q}$'s. Then by the proof above

$$\psi'' = \Psi' + T'_{\dot{q}-\dot{q}}.$$

And ψ'' is the kinetic energy actually assumed in the new system.

And we have proved that Ψ' is greater than $T_{\dot{q}}$ the kinetic energy of the old system.

Therefore *à fortiori* the kinetic energy in the new exceeds that in the original system.

In other words, if the masses be diminished, the kinetic energy will be increased by the sum of two quantities, the first being the amount by which the kinetic energy is diminished when the masses are diminished with unaltered velocities, the second being the kinetic energy in the new system with velocities equal to the difference of the old and new velocities.

20.] *If a material system at rest in any given configuration be set in motion in such a manner that the r velocities* $\dot{q}_1, \dot{q}_2 \ldots \dot{q}_r$ *have certain given values, and if the impressed impulses be such that* $p_{r+1}, p_{r+2} \ldots p_n$ *are separately zero, then the resulting motion of the system will be such as to render the kinetic energy the least possible consistent with the given velocities* $\dot{q}_1, \dot{q}_2 \ldots \dot{q}_r$. (Thomson's Theorem.)

For the conditions that the kinetic energy should be a maximum or minimum consistent with the r velocities $\dot{q}_1, \dot{q}_2 \ldots \dot{q}_r$ are

$$\frac{dT_{\dot{q}}}{d\dot{q}_{r+1}} = \frac{dT_{\dot{q}}}{d\dot{q}_{r+2}} = \ldots \frac{dT_{\dot{q}}}{d\dot{q}_n} = 0;$$

or $$p_{r+1} = p_{r+2} \ldots = p_n = 0.$$

Let $\dot{q}_{r+1}, \dot{q}_{r+2} \ldots \dot{q}_n$ be the values of the $n-r$ unknown velocities determined from these equations, and let T_0 be the value of the kinetic energy with these velocities, and let T_1 be its value when any other values as $\dot{q}_{r+1} + \dot{q}'_{r+1} \ldots \dot{q}_n + \dot{q}'_n$ are substituted for these $n-r$ velocities, the first r velocities remaining the same as before, then

$$T_1 = T_0 + T_{\dot{q}'} + \Sigma p \dot{q}', \text{ by Prop. VII.}$$

where $T_{\dot{q}'}$ is the value of T with the first r velocities separately zero and the last $n-r$ velocities $\dot{q}'_{r+1} \dots \dot{q}'_n$ respectively.

Also in $\Sigma p\dot{q}'$ the first r velocities are separately zero and the last $n-r$ momenta are also zero;

$$\therefore \quad \Sigma p\dot{q}' = 0.$$

$$\therefore \quad T_1 = T_0 + T_{\dot{q}}';$$

or

$$T_0 = T_1 - T_{\dot{q}}';$$

that is to say, T_0, the kinetic energy determined by the condition that the last $n-r$ momenta are separately zero, is less than the kinetic energy with momenta different from these, and the first r velocities the same as before, by the kinetic energy of the system in which each velocity is equal to the difference of the corresponding velocities in the original and altered system.

It follows from this Proposition that whenever a material system in any given configuration is set in motion by impulses entirely of given types in such a way that the velocities of the corresponding types have certain given values, then the motion of the system may be entirely determined by the condition that the kinetic energy assumed is the least possible with the given configuration and given velocities, the number of given equations among the velocities together with the equations of the form $p = 0$ being equal to the number of independent variables.

21.] Hence we may deduce the following theorem :—

If a material system at rest be set in motion by any impulses, the kinetic energy with which it moves off is the least which it can have consistently with the velocities assumed by the points at which the impulses are applied.

For suppose that the connections of the system are such that r of the generalised coordinates are known functions of the $3m$ coordinates x, y, z, &c. of the points of application of the impulses and of these variables only.

Let the points of application of the impulses be m in number, viz. $O_1, O_2 \dots O_m$; then the $3m$ coordinates of $O_1 \dots O_m$ are each of them determinate functions of the r coordinates $q_1, q_2, \dots q_r$.

Let X, Y, Z be rectangular components of any of the impulses acting at any one of these m points, then by definition any generalised component of impulse, as P_s, will be

$$\Sigma\left(X\frac{dx}{dq_s} + Y\frac{dy}{dq_s} + Z\frac{dz}{dq_s}\right).$$

Then P_s will be always zero unless s lie between 1 and r inclusive, for if x, y, z refer to any point of the system other than $O_1, O_2 \dots O_m$, the values of X, Y, Z are separately zero; and if x, y, z refer to one of these points, the values of $\frac{dx}{dq_s}, \frac{dy}{dq_s}, \frac{dz}{dq_s}$ will be separately zero if $s > r$, because the positions of these points are functions of the first r q's and of these only.

In this case therefore the components of impressed impulses $P_{r+1} \dots P_n$ are separately zero, and therefore the generalised components of momentum $p_{r+1} \dots p_n$ are separately zero; and therefore if the velocities of the m points, and consequently the values of $\dot{q}_1 \dots \dot{q}_r$, are given, the proposition of Art. 20 shews that the kinetic energy must be a minimum with these given velocities.

It may also happen that some of the momenta corresponding to $q_1 \dots q_r$—e.g. p_r and p_{r-1}—are zero. In that case the kinetic energy is not only the least possible consistently with $\dot{q}_1 \dots \dot{q}_r$, but also the least possible consistently with $\dot{q}_1 \dots \dot{q}_{r-2}$.

22.] The following are examples of the use of these theorems:

Example 1. The system of pulleys described in Art. 8 being at rest, let any velocity $\dot{q}_1$ in a vertical direction be given to the weight m_1 by an impulse applied at m_1: it is required to determine the initial motion of the system. If there be only one moveable pulley, we have only to make $p_2 = 0$, that is

$$(m_3 - m_2)\dot{q}_1 + (m_3 + m_2)\dot{q}_2 = 0,$$

or

$$\dot{q}_2 = \frac{m_2 - m_3}{m_2 + m_3}\dot{q}_1,$$

which determines the motion. In like manner if there be λ moveable pulleys, the expressions for p_2, p_3, &c. given in Art. 8, equated to zero, give as many linear equations as are necessary for determining $\dot{q}_2 \dots \dot{q}_{\lambda+1}$.

Example 2. In the chain of λ links discussed in Art. 8 let any velocities in the plane of the chain be given impulsively to P, the extremity of the rth link from O by impulses applied at P.

Here, unless r be unity, the system would lose generally two degrees of freedom if P were fixed, and therefore the rectangular coordinates of P might be expressed as functions of two generalised coordinates. In the system of coordinates employed in Art. 8 they are not in fact so expressed unless $r = 2$. Generally,

$$x = \Sigma^r{}_1\, a \cos\theta, \quad y = \Sigma^r{}_1\, a \sin\theta.$$

In order to determine the initial motion when $r > 2$, we must either first transform the coordinates, or seek by the general method of the calculus of variations to make T a minimum consistently with the given velocities of P, that is with

$$\Sigma^r{}_1\, a \sin\theta\,\dot{\theta} \quad \text{and} \quad \Sigma^r{}_1\, a \cos\theta\,\dot{\theta}.$$

If for instance the velocity of P in one direction only be given, and be produced by impulses acting in that direction only, we may take the given direction for axis of x, and then we have from p_1 to p_r inclusive

$$p \propto \sin\theta, \quad \text{and} \quad p_{r+1} = 0 \ldots p_\lambda = 0,$$

from which the p's may be determined as in Art. 10.

If the velocity of P be given in both directions, or if more than one point be struck, the expressions would assume complicated forms.

23.] Certain very interesting examples of the use of the propositions of Arts. 20 and 21 are given in Thomson and Tait's *Natural Philosophy*. These will repay fuller discussion.

For instance, a rigid body is set in motion by a blow applied at a certain point in such a way that the velocity of that point has a certain determinate value in magnitude and direction.

It is clear, from what we have just now proved, that all we have to do is to express the kinetic energy of the body in terms of the three component velocities of the point struck and three other variables, and to make this kinetic energy a minimum.

Let the body be referred to the principal axes through O, the point struck. Let u, v, w be the given velocities of O; $\omega_x, \omega_y, \omega_z$ the angular velocities round the coordinate axes; A, B, C the moments of inertia round the axes; $\bar{x}, \bar{y}, \bar{z}$ the coordinates of the centre of gravity; M the mass of the body. Then we have, since the component velocities of any element m of the mass of the body situated at x, y, z are

$$u+z\omega_y-y\omega_z, \quad v+x\omega_z-z\omega_x, \quad \text{and} \quad w+y\omega_x-x\omega_y,$$

$$2T = \Sigma m\left\{\{u+z\omega_y-y\omega_z\}^2+\{v+x\omega_z-z\omega_x\}^2+\{w+y\omega_x-x\omega_y\}^2\right\},$$

$$= \Sigma m(u^2+v^2+w^2)+A\omega_x^2+B\omega_y^2+C\omega_z^2$$

$$+\, 2M\left\{\bar{x}\{v\omega_z-w\omega_y\}+\bar{y}\{w\omega_x-u\omega_z\}+\bar{z}\{u\omega_y-v\omega_x\}\right\};$$

whence we obtain by the ordinary method, making T a minimum,

$$A\omega_x+M\{w\bar{y}-v\bar{z}\}=0,$$
$$B\omega_y+M\{u\bar{z}-w\bar{x}\}=0,$$
$$C\omega_z+M\{v\bar{x}-u\bar{y}\}=0,$$

which determine $\omega_x, \omega_y, \omega_z$. We might in this case obtain the same result from the assumption that the moment of momentum round each axis through O is zero.

24.] Again, an inextensible string is set in motion by impulses applied at its ends in such a way that the velocities assumed by the ends have certain given values.

We have to express the kinetic energy of the whole string, paying regard to the equation of continuity which expresses the inextensibility of the string, and remembering to take account of the given velocities of the points pulled.

This example is fully worked out by Thomson and Tait, pp. 226–229. We here vary the geometric treatment by introducing the notation of quaternions.

It is obvious that the terminal impulses are necessarily tangential, since any impulses applied at right angles to the tangent would generate in the extremity of the string an infinite velocity, without instantaneously affecting any other portions of the string.

Let $\mu\, ds$ be the mass of an element ds of the string at P.

Let ρ_P be the vector from the origin to P, $\dot{\rho}_P$ the vector velocity of P.

Also let ρ_Q, $\dot{\rho}_Q$ be the corresponding vectors for a neighbouring point Q of the string. Then, by the condition of inextensibility, $T(\rho_P - \rho_Q)$ is constant; that is,

$$\frac{d}{dt}T(\rho_P - \rho_Q) = 0,$$

$$\text{or} \quad S.(\rho_P - \rho_Q)(\dot{\rho}_P - \dot{\rho}_Q) = 0.$$

That is,
$$S.\frac{d\rho}{ds}\frac{d\dot{\rho}}{ds} = 0;$$

or, writing as usual ρ' for $\frac{d\rho}{ds}$,

$$S\rho'\frac{d\dot{\rho}}{ds} = 0. \quad \ldots \ldots \ldots \ldots \ldots \ldots \ldots \ldots \ldots \quad (1)$$

Again,
$$2T = \int_0^s \mu\dot{\rho}^2\,ds,$$

s being the length of the string. Then, making T a minimum subject to (1), we have by the calculus of variations,

$$\int_0^s S\left\{\mu\dot{\rho}\,\delta\dot{\rho} + \lambda\rho'\frac{d\,\partial\dot{\rho}}{ds}\right\}ds = 0,$$

λ being an indeterminate multiplier.

Integrating the second term by parts, and reducing to zero the terminal values of $\partial\dot{\rho}$, we obtain

$$\int_0^s S\left\{\mu\dot{\rho}\,\partial\dot{\rho} - \partial\dot{\rho}\frac{d}{ds}(\lambda\rho')\right\}ds = 0,$$

$$\text{whence} \quad \mu\dot{\rho} = \frac{d}{ds}(\lambda\rho'),$$

$$\text{or} \quad \dot{\rho} = \frac{1}{\mu}\frac{d}{ds}(\lambda\rho'), \quad \ldots \ldots \ldots \ldots \ldots \ldots \ldots \quad (2)$$

and
$$\frac{d\dot{\rho}}{ds} = \frac{d\frac{1}{\mu}}{ds}\frac{d\lambda}{ds}\rho' + \frac{d\frac{1}{\mu}}{ds}\lambda\rho'' + \frac{1}{\mu}\frac{d^2\lambda}{ds^2}\rho' + \frac{2}{\mu}\frac{d\lambda}{ds}\rho'' + \frac{\lambda}{\mu}\rho'''.$$

We then substitute this value of $\frac{d\dot{\rho}}{ds}$ in the equation (1),

$$S\rho'\frac{d\dot{\rho}}{ds} = 0,$$

attending to the following known relations,

$$S\rho'^2 \text{ i.e. } \rho'^2 = -1, \quad S\rho'\rho'' = 0, \quad S\rho'\rho''' = -\rho''^2 = \frac{1}{r^2},$$

where r is the radius of absolute curvature, and obtain

$$\frac{1}{\mu}\frac{d^2\lambda}{ds^2}+\frac{d\frac{1}{\mu}}{ds}\frac{d\lambda}{ds}-\frac{\lambda}{\mu r^2}=0. \quad \ldots\ldots\ldots\ldots \quad (3)$$

This determines λ with two arbitrary constants.

Again, we have from (2)

$$\dot\rho=\frac{1}{\mu}\frac{d\lambda}{ds}\rho'+\frac{\lambda}{\mu}\rho''.$$

Now ρ' is a unit vector in direction of the tangent, and ρ'' a vector proportional to $\frac{1}{r}$ in direction towards the centre of curvature. Hence we see that λ is the impulsive tension at a point in the string, and that the velocity which the point acquires instantaneously is the resultant of $\frac{1}{\mu}\frac{d\lambda}{ds}$ tangential, and $\frac{\lambda}{\mu r}$ towards the centre of curvature, and is independent of any variation in the plane of curvature. At either end ρ'' is indeterminate, therefore (2) gives at each end only one equation for determining the two arbitrary constants, namely that obtained by equating $\frac{1}{\mu}\frac{d\lambda}{ds}$ to the given tangential velocity at the end in question. If the velocity at one end only be given, then at the other end, as we cannot make $\partial\rho$ zero, we must, in order for T to be minimum, make λ zero. $\lambda=0$ is then one of the two equations for determining the constants.

25.] Again, a smooth vessel, full of incompressible fluid, is set in motion with any given velocity, find the resulting motion of the fluid.

In this case the position of each point of the rigid body represented by the containing vessel is a determinate function of six of the independent variables, and the given velocities of all the points of the vessel also determine the six component velocities corresponding to these six variables. Also by whatever impulses the vessel is set in motion these impulses must pass through some points of the vessel. Therefore the general proposition of Art. 21 applies, and we have only to express the kinetic energy of the vessel and contained fluid, regard being paid to the equation of continuity of the fluid and to the velocity conditions of the vessel, and to make this kinetic energy a minimum.

Let θ be the velocity given to any point P on the surface of the vessel resolved in the normal to the surface. Then a particle of the fluid adjacent to P takes the same normal velocity θ.

Let $\delta\nu$ be an element of the normal. Let K be the density of the fluid at any point. K is therefore an essentially positive quantity. Let x, y, z be the coordinates of any point referred to rectangular axes. Let V be a function of x, y, z satisfying the following conditions, viz.

$$\frac{dV}{d\nu} = \theta$$

at every point on the surface, and

$$\frac{d}{dx}\left(K\frac{dV}{dx}\right) + \frac{d}{dy}\left(K\frac{dV}{dy}\right) + \frac{d}{dz}\left(K\frac{dV}{dz}\right) = 0$$

at every point within the vessel *.

Let u, v, w be the initial velocities taken by a particle of the fluid. Then a motion in which

$$u = \frac{dV}{dx}, \quad v = \frac{dV}{dy}, \quad \text{and} \quad w = \frac{dV}{dz},$$

satisfies the surface condition

$$\frac{dV}{d\nu} = \theta \quad \dots\dots\dots\dots \quad (1)$$

at every point on the surface, and also satisfies the equation of continuity, viz.

$$\frac{d}{dx}(Ku) + \frac{d}{dy}(Kv) + \frac{d}{dz}(Kw) = 0,$$

$$\text{or} \quad \frac{d}{dx}\left(K\frac{dV}{dx}\right) + \frac{d}{dy}\left(K\frac{dV}{dy}\right) + \frac{d}{dz}\left(K\frac{dV}{dz}\right) = 0 \quad \dots\dots \quad (2)$$

at every point within the vessel, and is therefore a possible motion of the liquid subject to the given surface conditions.

If we can show that it has less kinetic energy than any other motion satisfying the same conditions, it must by our principal proposition be the motion actually assumed by the liquid.

* See Maxwell's *Electricity*, vol. i. p. 104.

If the actual velocities be not

$$\frac{dV}{dx},\quad \frac{dV}{dy},\quad \text{and}\quad \frac{dV}{dz}$$

at every point, let them be

$$\frac{dV}{dx}+a,\quad \frac{dV}{dy}+\beta,\ \text{and}\ \frac{dV}{dz}+\gamma.$$

Then in order that the surface conditions and the equation of continuity may be satisfied, we must have

$$a=0,\quad \beta=0,\quad \gamma=0$$

at every point on the surface, and

$$\frac{d}{dx}(Ka)+\frac{d}{dy}(K\beta)+\frac{d}{dz}(K\gamma)=0$$

at every point within the vessel.

Then the kinetic energy of the motion $\frac{dV}{dx}+a$, &c. is

$$\frac{1}{2}\iiint K\left\{\left(\frac{dV}{dx}\right)^2+\left(\frac{dV}{dy}\right)^2+\left(\frac{dV}{dz}\right)^2\right\}dx\,dy\,dz$$

$$+\frac{1}{2}\iiint K\{a^2+\beta^2+\gamma^2\}\,dx\,dy\,dz$$

$$+\iiint K\left\{a\frac{dV}{dx}+\beta\frac{dV}{dy}+\gamma\frac{dV}{dz}\right\}dx\,dy\,dz.$$

By Green's theorem the third line is equal to

$$\iint KVa\,dy\,dz+\iint KV\beta\,dx\,dz+\iint KV\gamma\,dx\,dy$$

$$-\iiint V\left\{\frac{d}{dx}(Ka)+\frac{d}{dy}(K\beta)+\frac{d}{dz}(K\gamma)\right\}dx\,dy\,dz,$$

and is therefore zero; since a, β, γ are zero on the surface, and the quantity under the triple integral is zero within the vessel. Hence

$$\iiint K\left\{\left(\frac{dV}{dx}\right)^2+\left(\frac{dV}{dy}\right)^2+\left(\frac{dV}{dz}\right)^2\right\}dx\,dy\,dz$$

is less than

$$\iiint K\left\{\left(\frac{dV}{dx}+a\right)^2+\left(\frac{dV}{dy}+\beta\right)^2+\left(\frac{dV}{dz}+\gamma\right)^2\right\}dx\,dy\,dz,$$

and therefore the kinetic energy of the motion $\frac{dV}{dx}$, &c. is less than that of any other motion satisfying the given surface conditions and the equation of continuity. This motion is therefore the actual motion. The process itself shews that there can be only one function, V, of x, y, z satisfying conditions (1) and (2), except as such function may be varied by the addition of a constant. Therefore $\frac{dV}{dx}$, &c., or u, v, and w, have single values at every point in the fluid. In other words, for any given initial motion of the containing vessel there is a single determinate motion of the fluid.

Evidently $\frac{du}{dy} = \frac{dv}{dx}$, &c., and the motion is of the kind called *non-rotational*. V is called the *velocity potential*.

The above investigation would evidently apply if, instead of a single vessel enclosing the liquid, there were several vessels, and if the liquid had immersed in it any rigid or flexible bodies bounded by closed surfaces.

26.] In Thomson and Tait's *Natural Philosophy* the use of generalised coordinates is illustrated in a very interesting manner by their application to certain cases of fluid motion.

Given an incompressible homogeneous fluid, either infinite in extent or bounded by any finite closed surfaces of any form, and with any rigid or flexible bodies moving through it, it may be proved that the kinetic energy of the whole fluid is known at any instant if the velocities of the containing surfaces and those of the moving bodies are known.

This truth can be established by some such reasoning as follows. It is true that although the positions of the containing surfaces and immersed bodies be known, the system has in respect of the relative motions of the particles of the fluid a practically infinite number of degrees of freedom left, and might conceivably have kinetic energy although the containing surfaces and immersed bodies were all at rest, yet we may suppose the relative positions of all the particles of the fluid to be determined by certain generalised coordinates $q_1 \ldots q_r$, r being sufficiently great, and $q_{r+1} \ldots q_n$ being the remaining coordinates of the system, those namely which define the position of the containing surfaces and immersed bodies. If now the containing surfaces or immersed bodies be set in motion by any impulses from rest, we have already seen that the kinetic energy of the whole system is the least which is consistent with the velocities assumed by those surfaces and bodies, that is with $\dot{q}_{r+1} \ldots \dot{q}_n$, and therefore, by Art. 20, the generalised components of momentum corresponding to $q_1 \ldots q_r$ are severally zero. It is evident also that any impulses which if applied to the system at rest would make $p_1 \ldots p_r$ zero, will not if applied to the system in motion, however they may alter the velocities, give to $p_1 \ldots p_r$ any values. Hence, so far as impulses applied to the surfaces and immersed bodies are concerned, $p_1 \ldots p_r$ remain zero for all time.

If any finite forces act on the system, the same result as regards $p_1 \ldots p_r$ follows from Lagrange's equations. For

$$\frac{dp}{dt} = \frac{dT}{dq} - \frac{dU}{dq}$$

for each of the coordinates $q_1 \ldots q_r$. Now if the containing surfaces and immersed bodies were all at rest and fixed in space the forces acting on the system could have no tendency to produce relative motion among the particles of the fluid, it being homogeneous. Hence for $q_1 \ldots q_r$, $\frac{dU}{dq} = 0$; and evidently also $\frac{dT}{dq} = 0$. Hence $\frac{dp}{dt} = 0$; and the motion being from rest, $p = 0$.

It follows that in such a system as we have supposed, to whatever finite or impulsive forces it may be subject, provided the impulses act at points in the containing surfaces or immersed bodies, the components of momentum $p_1 \ldots p_r$ are always zero. And therefore the kinetic energy of the entire system at any instant can be expressed in terms of the momenta corresponding to the remaining coordinates $q_{r+1} \ldots q_n$, which define the positions of the containing surfaces and immersed bodies.

It follows, by Arts. 20 and 25, that the motion of the entire system at any instant is that which it would take if, the whole being at rest, their actual velocities at that instant were impulsively given to the containing surfaces and immersed bodies. If, therefore, the positions of these containing or immersed surfaces are determined by a certain finite number of coordinates, the whole motion of the fluid and of the immersed bodies may also be determined in terms of these coordinates.

27.] The first two of the following three examples are taken from Thomson and Tait, p. 262, &c.

Example 1. A ball is set in motion through a mass of frictionless incompressible fluid extending infinitely in all directions on one side of an infinite plane and originally at rest.

The position of the ball, and therefore by our general proposition, the whole motion is determined if the coordinates of the ball's centre x, y, z at the time t are known.

Let the axis of x be taken perpendicular to the bounding plane through any point whatever of that plane; then the kinetic energy T must be a quadratic function of $\dot{x}$, $\dot{y}$, and $\dot{z}$, with coefficients certain functions of x, y, and z.

It is clear that T remains of the same value when either $\dot{y}$ or $\dot{z}$ has its sign reversed, and therefore the terms in $\dot{x}\dot{y}$, $\dot{x}\dot{z}$, and $\dot{y}\dot{z}$ do not occur in T, which is therefore reduced to the form

$$\tfrac{1}{2}\{P\dot{x}^2 + Q\dot{y}^2 + R\dot{z}^2\},$$

where P, Q, and R are functions of x only.

From the symmetry it is clear that $Q = R$, and hence

$$T = \tfrac{1}{2}\{P\dot{x}^2 + Q(\dot{y}^2 + \dot{z}^2)\}.$$

If therefore X, Y, and Z are the generalised components of force corresponding to x, y, and z, Lagrange's equations give us

$$P\ddot{x} + \tfrac{1}{2}\left\{\frac{dP}{dx}\dot{x}^2 - \frac{dQ}{dx}(\dot{y}^2 + \dot{z}^2)\right\} = X,$$

$$Q\ddot{y} + \frac{dQ}{dx}\dot{y}\dot{x} = Y, \qquad Q\ddot{z} + \frac{dQ}{dx}\dot{z}\dot{x} = Z.$$

Example 2. A solid of revolution moving through a frictionless incompressible fluid infinitely extended so as to keep its axis always in one plane.

In this case there are three degrees of freedom, and therefore three independent velocities in terms of which the whole motion may be determined.

Let these be chosen as the two components of the velocity of any point in the axis of figure, and the angular velocity about an axis through the same point perpendicular to the plane in which that axis moves. It is assumed that the body has no rotation about its axis of figure.

If u and q be the resolved parts of the velocity of the point along and perpendicular to the axis of figure, and w the angular velocity about the axis through this point, it is clear from our general proposition that the whole kinetic energy of the body and fluid is

$$\tfrac{1}{2}\{Au^2+Bq^2+Cw^2+Dwq\}.$$

For the reversal of the sign of u cannot affect T.

A, B, C, D also are obviously constants, since the liquid is of infinite extent.

By properly selecting the aforesaid point in the axis the equation for T may be reduced by obvious reductions to

$$T=\tfrac{1}{2}\{Au^2+Bv^2+Ew^2\};$$

u and v being the velocities of the new point in the aforesaid directions.

If θ be the angle between the axis of figure and the axis of x, and x and y the coordinates of the aforesaid point, we get

$$w=\dot{\theta},\quad u=\dot{x}\cos\theta+\dot{y}\sin\theta,\quad v=\dot{y}\cos\theta-\dot{x}\sin\theta;$$

$$\frac{dT}{d\dot{\theta}}=E\dot{\theta},\quad \frac{dT}{d\dot{x}}=Au\cos\theta-Bv\sin\theta,\quad \frac{dT}{d\dot{y}}=Au\sin\theta+Bv\cos\theta;$$

$$\frac{dT}{d\theta}=-(A-B)uv,\quad \frac{dT}{dx}=0,\quad \frac{dT}{dy}=0.$$

Also, if λ, ξ, η be the generalised components of impulse corresponding to w, x, and y,

$$\lambda=\frac{dT}{d\dot{\theta}}=\mu\dot{\theta},$$

$$\xi=Au\cos\theta-Bv\sin\theta,$$

$$\eta=Au\sin\theta+Bv\cos\theta.$$

And Lagrange's equations give us,

$$E\frac{d^2\theta}{dt^2}+\frac{A-B}{2AB}\{(\xi^2-\eta^2)\sin 2\theta-2\xi\eta\cos 2\theta\}=L,$$

$$\frac{d\xi}{dt}=X,\quad \frac{d\eta}{dt}=Y;$$

L, X, and Y being generalised components of force corresponding to w, x, and y.

If L, X, and Y each $= 0$, and if the axes be so taken that $\eta = 0$, as is clearly possible in this case, we get

$$E\frac{d^2\theta}{dt^2} + \frac{A-B}{2AB}\xi^2 \sin 2\theta = 0,$$

or $$E\frac{d^2\phi}{dt^2} + (\frac{A-B}{AB}\xi^2)\sin\phi = 0.$$

The case of the common pendulum where $\phi = \frac{\theta}{2}$.

Example 3. As a third illustration we may take a case of motion in three dimensions as follows.

An ellipsoid of revolution moving in an infinite mass of frictionless incompressible fluid—no forces.

If u, v, w be the velocities of the centre resolved parallel to the three principal axes, and if ω_1, ω_2, ω_3 be the angular velocities about these axes, it is clear from our general proposition that the whole kinetic energy, T, of the liquid and ellipsoid may be expressed as a quadratic function of these six quantities u, v, w, ω_1, ω_2, ω_3.

Also from the perfect symmetry it is clear that terms involving the products of these quantities cannot appear in T, and therefore that

$$T = \tfrac{1}{2}\{Au^2 + Av^2 + Bw^2 + D\omega_1^2 + D\omega_2^2 + C\omega_3^2\};$$

where A, B, and D will be certain constant quantities, and where C is the moment of inertia of the ellipsoid about its axis.

If x, y, z be the rectangular coordinates of the centre, and if θ, ϕ, ψ be the angular coordinates of ordinary use in determining the orientation of a rigid body, we get by obvious substitution and reduction,

$$\frac{dT}{d\dot{x}} = Au\{\cos\psi\cos\phi\cos\theta - \sin\phi\sin\psi\}$$
$$- Av\{\cos\psi\sin\phi\cos\theta + \cos\phi\sin\psi\} + Bw\cos\psi\sin\theta,$$

$$\frac{dT}{d\dot{y}} = Au\{\sin\psi\cos\phi\cos\theta + \cos\psi\sin\phi\}$$
$$+ Av\{\cos\phi\cos\psi - \sin\psi\sin\phi\cos\theta\} + Bw\sin\psi\sin\theta,$$

$$\frac{dT}{d\dot{z}} = Au\sin\theta\cos\phi - Av\sin\theta\sin\phi + Bw\cos\theta.$$

If the motion be produced from rest by an impressed impulse F parallel to the axis of z, then

$$\frac{dT}{d\dot{x}} = 0, \quad \frac{dT}{d\dot{y}} = 0, \quad \frac{dT}{d\dot{z}} = F;$$

Now $$\frac{dT}{d\dot{x}} = 0 \text{ and } \frac{dT}{d\dot{y}} = 0$$

give $$\frac{u}{\cos\phi} = -\frac{v}{\sin\phi} = -\frac{B}{A}\tan\theta\, w,$$

or $$u = -\frac{B}{A}\tan\theta\cos\phi\, w, \quad v = +\frac{B}{A}\tan\theta\sin\phi\, w;$$

and substituting in $\frac{dT}{d\dot{z}} = F$, we get

$$w = \frac{F}{B}\cos\theta;$$

whence substituting in the equations, giving $\dot{x}$, $\dot{y}$, $\dot{z}$ in terms of u, v, and w, we get

$$\dot{x} = F\left(\frac{1}{A} - \frac{1}{B}\right)\cos\theta\sin\theta\cos\psi, \quad \dot{y} = F\left(\frac{1}{A} - \frac{1}{B}\right)\sin\theta\cos\theta\sin\psi,$$

$$\dot{z} = F\left(\frac{\sin^2\theta}{A} + \frac{\cos^2\theta}{B}\right).$$

Since

$$\omega_3 = \dot{\phi} + \dot{\psi}\cos\theta, \quad \omega_2 = \dot{\theta}\cos\phi - \dot{\psi}\sin\theta\sin\phi,$$

$$\omega_1 = \dot{\theta}\sin\phi + \dot{\psi}\sin\theta\cos\phi.$$

It follows that T is independent of ψ, and therefore Lagrange's equation corresponding to the coordinate ψ becomes

$$\frac{d}{dt}\left(\frac{dT}{d\dot{\psi}}\right) = 0, \text{ or } \frac{dT}{d\dot{\psi}} = \text{E (a constant)}.$$

But $$\frac{dT}{d\dot{\psi}} = D\omega_1\frac{d\omega_1}{d\dot{\psi}} + D\omega^2\frac{d\omega_2}{d\dot{\psi}} + C\omega_3\frac{d\omega_3}{d\dot{\psi}}$$

$$= D\sin\theta\,\{\omega_1\cos\phi - \omega_2\sin\phi\} + C\omega_3\cos\theta$$

$$= D\sin^2\theta\dot{\psi} + C\omega_3\cos\theta;$$

$$\therefore \quad D\sin^2\theta\psi + C\omega_3\cos\theta = E.$$

And T may be reduced to

$$\tfrac{1}{2}D\dot{\theta}^2 + \tfrac{1}{2}D\sin^2\theta\dot{\psi}^2 + \tfrac{1}{2}C\omega_3^2 + \tfrac{1}{2}F^2\left(\frac{\sin^2\theta}{A} + \frac{\cos^2\theta}{B}\right).$$

28.] Lord Rayleigh has pointed out a remarkable analogy between the dynamical theorems hitherto demonstrated and certain statical theorems, generalised components of velocity being replaced by small displacements, generalised components of momentum by impressed forces, and kinetic energy by potential energy of deformation.

For example, suppose a statical system under the influence of given impressed forces, which are either constant or functions of the positions of the particles, to be in a position of stable equilibrium; the potential energy of the system must then be a minimum. Let it be V_0. Let the system be slightly displaced by the application of certain additional forces, and let the generalised coordinates of the displaced position reckoned from the position of stable equilibrium be $q_1 \dots q_n$. The potential energy of the displaced system will then be $V_0 + V$, where V is a quadratic function of $q_1 \dots q_n$, involving generally coefficients functions of the coordinates, $a, b, c, \dots$, of the position of stable equilibrium. V is defined to be the *potential energy of deformation.*

Also the generalised components of force $P_1 \dots P_n$, required to produce the deformation $q_1 \dots q_n$, are linear functions of $q_1 \dots q_n$ with coefficients functions of $a, b, c, \dots$ such that

$$P_1 = \frac{dV}{dq_1}, \quad P_2 = \frac{dV}{dq_2}, \text{ \&c.; and } V = \tfrac{1}{2}\Sigma Pq.$$

We can then by means of these linear equations prove a series of propositions exactly analogous to Propositions IV–VII of Art. 9; and in particular we can prove that if $P_1 \dots P_n$ be the forces producing the deformation $q_1 \dots q_n$, while $P_1' \dots P_n'$ produce $q_1' \dots q_n'$ from the same position of stable equilibrium, then

$$\Sigma P q' = \Sigma P' q, \quad \text{or} \quad \Sigma \frac{dV_q}{dq} q' = \Sigma \frac{dV_{q'}}{dq'} q.$$

Then we may prove a proposition analogous to that of the maximum kinetic energy (Art. 16) above, namely, that if such a material system be held in equilibrium, in any position slightly displaced from that of stable equilibrium, by means of forces applied from without, the potential energy of such displacement will be greater the greater the number of degrees of freedom, and that if the system be subject to any constraints, and so constrained be held in equilibrium in a position slightly displaced from the original position of stable equilibrium by means of the same external forces as before, the potential energy of the free system in its displaced position will be greater than that of the constrained system in its displaced position by the potential energy of the difference of displacements in the two displaced positions.

Let $q_1 \dots q_n$ be the displacements in the displaced position of the free system, reckoned from the position of stable equilibrium, and let V_q be the potential energy of displacement in this case. Let $q_1' \dots q_n'$ and $V_{q'}$ be the corresponding quantities in the constrained system. Let $P_1 \dots P_n$ be the external or additional impressed forces in both cases. Then we have, as above stated,

$$\Sigma \frac{dV_q}{dq} q' = \Sigma \frac{dV_{q'}}{dq'} q,$$

and
$$P_1 = \frac{dV_q}{dq_1}, \quad P_2 = \frac{dV_q}{dq_2}, \text{ \&c.}$$

In the constrained system it will no longer be true that

$$P_1 = \frac{dV_{q'}}{dq_1'}, \quad P_2 = \frac{dV_{q'}}{dq_2'}, \text{ \&c.},$$

because the displacements are no longer independent, but, by reasoning in all respects analogous to that of Art. 16 above, we must have, by the principle of virtual velocities,

$$\Sigma \left(P - \frac{dV_{q'}}{dq'}\right) q' = 0,$$

or
$$\Sigma \left(\frac{dV_q}{dq} - \frac{dV_{q'}}{dq'}\right) q' = 0. \qquad (1)$$

Also in the constrained system,

$$V_{q'} = \tfrac{1}{2} \Sigma \frac{dV_{q'}}{dq'} q';$$

therefore

$$V_q - V_{q'} = \tfrac{1}{2} \Sigma \frac{dV}{dq} q - \tfrac{1}{2} \Sigma \frac{dV_{q'}}{dq'} q'$$

$$= \tfrac{1}{2} \Sigma \frac{dV_q}{dq} q - \tfrac{1}{2} \Sigma \frac{dV_q}{dq} q' \text{ by (1)}$$

$$= \tfrac{1}{2} \Sigma \frac{dV_q}{dq} (q - q').$$

But by (1) $$\Sigma \frac{dV_{q'}}{dq'} q' = \Sigma \frac{dV_q}{dq} q' = \Sigma \frac{dV_{q'}}{dq'} q;$$

$$\therefore \quad \Sigma \frac{dV_{q'}}{dq'}(q-q') = 0;$$

$$\therefore \quad \tfrac{1}{2} \Sigma \frac{dV_q}{dq}(q-q') = \tfrac{1}{2}\Sigma\left(\frac{dV_q}{dq} - \frac{dV_{q'}}{dq'}\right)(q-q'),$$

$$= V_{q-q'}. \text{ (See Prop. VII. Art. 9.)}$$

That is, $$V_q - V_{q'} = V_{q-q'},$$

whence the proposition is proved.

29.] Again, if the expression $\Sigma PQ - V_Q$ be denoted by Ψ, where the P's are the given forces, and the Q's any whatever small displacements, and V_Q the potential energy of deformation corresponding to the Q's, and if ψ and ψ' be the values of Ψ when q and q' respectively are substituted for Q, that is $\psi = V_q$, $\psi' = V_{q'}$, then we may easily prove, as in Art. 18 above, *mutatis mutandis*, that

$$\psi - \Psi = \tfrac{1}{2}\Sigma.\left(\frac{dV_q}{dq} - \frac{dV_Q}{dQ}\right)(q-Q);$$

from which follows, as a particular case, the result already obtained,

$$V_q - V_{q'} = \psi - \psi' = \tfrac{1}{2}\Sigma.\left(\frac{dV_q}{dq} - \frac{dV_{q'}}{dq'}\right)(q-q') = V_{q-q'}.$$

30.] Hence we may shew that if the stiffness in any part or parts of the system be diminished, the connexions remaining unchanged, the potential energy of deformation will be increased. For if the displacements were the same it is evident that the potential energy would be diminished, there being less stiffness, that is, $V_q' < V_q$, if V_q' be the potential energy of deformation in the new, V_q in the original system with the same displacements $q_1 \dots q_n$.

Now in the function Ψ, formed for the new system, let the Q's be the original q's, and let Ψ' be the value of ψ in this case.

Then $$\Psi' = \Sigma Pq - V_q';$$

$$\therefore \quad \Psi' > \Sigma Pq - V_q \text{ because } V_q' < V_q;$$

$$\therefore \quad \Psi' > V_q, \text{ since } \Sigma Pq - V_q = V_q.$$

But if $q_1' \dots q_n'$ be the actual values of the displacements in the position of equilibrium of the new system under the impressed forces P, it follows, as above proved, that $V_{q'}' > \Psi'$; therefore, *à fortiori*, $V_{q'}' > V_q$.

31.] There is also a statical analogue to the theorem of minimum kinetic energy of Art. 20, which may be stated as follows :—

If a material system be held in a deformed position with given values of certain of the displacements, suppose $q_1 \dots q_r$, reckoned from the position of stable equilibrium, then the potential energy of deformation will be the least possible when the external or additional forces by which the displacements are produced are exclusively of the types corresponding to those displacements, and the potential energy of any other deformed position having the same values of $q_1 \dots q_r$ exceeds this least potential energy by the potential energy of the displacement which is the difference of the two positions.

The proof of this is analogous to that of the corresponding dynamical theorem. Let $P_1 \dots P_r$ be the forces necessary to produce the given displacements $q_1 \dots q_r$ when $P_{r+1} \dots P_n$ are severally zero. Let V be the potential energy of deformation in this case, V' that in some other deformed position having the same values of $q_1 \dots q_r$; and let $P_1 + P_1'$, $P_2 + P_2'$, &c. be the forces, and $q_1 + q_1'$, $q_2 + q_2'$, &c. the displacements in the latter case; then by hypothesis every q' from q_1' to q_r' inclusive is zero, and every P from P_{r+1} to P_n inclusive is zero; therefore

$$\Sigma P q' = \Sigma P' q = 0;$$

therefore

$$V' = \tfrac{1}{2} \Sigma (P + P')(q + q')$$
$$= \tfrac{1}{2} \Sigma P q + \tfrac{1}{2} \Sigma P' q' = V + \tfrac{1}{2} \Sigma P' q',$$

as was to be proved.

Hence we can deduce a theorem corresponding to that of Art. 21, viz.

If a material system in stable equilibrium under the action of its own forces undergo any small displacement or deformation by fresh forces applied from without, being so forced into a new position of equilibrium, the potential energy gained by such deformation is the least which the system can have consistently with the displacements, whatever they may be, *of the points at which the fresh forces are applied.*

32.] Our dynamical equations have also analogues in electrostatics. It can be shewn, for instance, that in any system of conductors in equilibrium relations exist analogous to those established for a dynamical system, generalised components of momentum and velocity being replaced by the potentials and charges of the several conductors, and kinetic energy by the intrinsic energy of the system, that is to say, the whole work which would have to be done to bring the charges from an infinite distance to the several conductors against their mutual repulsions. It is understood that the charges of the same sign repel one another according to the law of the inverse square.

Let $C_1 \dots C_s$ be the several conductors, $q_1 \dots q_n$ the generalised coordinates defining their positions in space, $e_1 \dots e_s$ their charges, $V_1 \dots V_s$ their potentials when the system is in equilibrium, and E the intrinsic energy. Then the work which would have to be done to bring an infinitely small quantity of electricity, de, to the conductor C_1 from an infinite distance is evidently $V_1 de$. Hence we obtain generally

$$\frac{dE}{de} = V.$$

Again, let us suppose all the charges to be originally zero, and to be gradually increased *pari passu* in the same ratio till they attain their value in the actual system; the potentials at any instant during this gradual variation are proportional to the charges at the instant. It follows, as shewn by Maxwell, *Electricity and Magnetism*, part I, chapter iii, that each potential is a linear function of all the charges, with coefficients depending on the forms of the conductors, and the coordinates $q_1 \dots q_n$ defining their positions in space.

It follows also that $\quad E = \frac{1}{2} \Sigma Ve;$ *

and E is therefore a quadratic function of the charges having coefficients functions of the q's.

If the charges $e_1 \dots e_s$ produce potentials $V_1 \dots V_s$, while $e_1' \dots e_s'$ produce $V_1' \dots V_s'$, evidently $e_1 - e_1'$, &c. will produce $V_1 - V_1'$, &c.

* For suppose the charges to be introduced uniformly during any time. Then after time t they will be $K_1 t \dots K_s t$, and the potentials $C_1 t \dots C_s t$, where the K's and C's are constants. Then, since $\frac{dE}{de} = V$, we have

$$\delta E = \Sigma V \delta e = \Sigma C K t \delta t; \text{ whence } E = \tfrac{1}{2} \Sigma C K t^2 = \tfrac{1}{2} \Sigma V e,$$

no constant of integration being required, because when $t = 0$ E $= 0$.

If the linear equations be

$$V_1 = A_{11}e_1 + A_{12}e_2 + \&c.,$$
$$V_2 = A_{21}e_1 + A_{22}e_2 + \&c.;$$

then, since $\frac{dE}{de} = V$, we must have $A_{12} = A_{21}$, &c., that is, generally,

$$\frac{dV_a}{de_b} = \frac{dV_b}{de_a}.$$

Conversely, every e is a linear function of all the V's, and E may be expressed as a quadratic function of the V's with coefficients functions of the q's. When so expressed we shall write it E_V, and when expressed as a quadratic function of the charges, E_e.

It follows then from the linear equations connecting V and e, that

$$\frac{dE_V}{dV} = e.$$

Also that if $V_1 \ldots V_s$ be the potentials of the several conductors when the charges are $e_1 \ldots e_s$, and if $V_1' \ldots V_s'$ be the potentials of the same system of conductors in the same positions when the charges are $e_1' \ldots e_s'$, then

$$\Sigma V e' = \Sigma V' e.$$

This result can be established by an independent method; see an article by Clausius in the *Philosophical Magazine*, vol. iv, Fifth Series, p. 454.

33.] We can now prove as in Proposition III of Art. 9 that

$$\frac{dE_e}{dq} + \frac{dE_V}{dq} = 0,$$

the potentials in the one coefficient, and the charges in the other, being treated as constant, and the forms of the conductors in either case unaltered.

For since $$E_e + E_V = 2E = \Sigma Ve,$$

let us suppose e, V, and q all to vary.

Then we have

$$\Sigma \frac{dE_e}{dq}\partial q + \Sigma \frac{dE_e}{de}\partial e + \Sigma \frac{dE_V}{dq}\partial q + \Sigma \frac{dE_V}{dV}\partial V = \Sigma(V\partial e + e\partial V),$$

the summations being for all ∂q's or all ∂e's &c. as the case may be.

But $$\frac{dE_e}{de} = V \text{ and } \frac{dE_V}{dV} = e;$$

therefore the above equation is reduced to

$$\Sigma \frac{dE_e}{dq} \partial q + \Sigma \frac{dE_V}{dq} \partial q = 0,$$

in which the summation is for all the ∂q's.

And since the ∂q's are independent, therefore for each coordinate q,

$$\frac{dE_e}{dq} + \frac{dE_V}{dq} = 0.$$

Now $\frac{dE_e}{dq} \partial q$ is the ~~diminution~~ increase of the intrinsic energy of the system consequent on the conductors undergoing the displacement in space denoted by ∂q, all the charges remaining unaltered; and therefore $-\frac{dE_e}{dq}$ ~~measures~~ is the mechanical force tending to displace the conductors in the manner denoted by ∂q.

Similarly $-\frac{dE_V}{dq}$ is the mechanical force tending to displace them in the same manner, if by any means the potentials be maintained constant during the displacement, while the charges vary. And the equation just obtained shows that the resultant mechanical forces are equal and opposite in the two cases. This result is obtained in a different way by Maxwell in the work above referred to, vol. i. p. 95.

34.] If any two or more conductors originally insulated be connected together, so as to form one conductor, they acquire of course uniform potential, and a new distribution of their charges takes place, the potentials of other parts of the system undergoing corresponding alterations. If $V_1 \ldots V_s$ and $e_1 \ldots e_s$ be the original potentials and charges, and $V_1' \ldots V_s'$ and $e_1' \ldots e_s'$ those after the connexion is established, we can prove the following theorem, viz.

$$\Sigma V e' = \Sigma V' e = \Sigma V' e' \text{ or } \Sigma (V - V') e' = 0$$

being the analogue of the equation

$$\Sigma (p - p') \dot{q}' = 0,$$

deduced in Art. 16 from D'Alembert's principle.

For in the case of every conductor which retains its insulation $e = e'$, and therefore $V'e = V'e'$.

In case of a group of conductors which become connected, V' is the same for all members of the group, and the sum of the charges is unaltered. Therefore

$$\Sigma V'e = \Sigma V'e',$$

Σ denoting summation for the group. It follows that for the entire system $\Sigma Ve' = \Sigma V'e = \Sigma V'e'$. .. .. (1)

Hence we can prove a theorem analogous to that of Art. 16, viz. that if any two or more conductors be connected so as to form one conductor, the intrinsic energy of the entire system is diminished by an amount equal to the intrinsic energy which the system would have, if the charge on each conductor in the entire system were the difference between its charges in the original and altered state; that is, that

$$\Sigma Ve = \Sigma V'e' + \Sigma (V - V')(e - e').$$

For $\Sigma Ve - \Sigma V'e' = \Sigma Ve - \Sigma Ve'$,

$$= \Sigma V(e - e'),$$

$$= \Sigma (V - V')(e - e');$$

because $\Sigma V'(e - e') = 0$ by (1);

i.e. $\Sigma Ve = \Sigma V'e' + \Sigma (V - V')(e - e')$,

as was to be proved.

The loss of intrinsic energy is therefore equal to the work which would have to be done to bring to all the conductors, supposed originally uncharged and insulated, the charges $e - e'$.

It follows that if a given quantity of electricity be distributed over a surface, the intrinsic energy is the least possible when the distribution is such as to make the potential uniform over the surface. And the same law holds for a number of surfaces if the charges on each be given.

Hence also if a number of insulated conductors be so charged as to have potentials $V_1 \ldots V_s$ respectively, then if they be all connected together, they will assume the common potential

$$V' = \frac{\Sigma V\sigma}{\Sigma \sigma},$$

where σ is for each conductor proportional to the charge which the conductor has after the connexion is established.

35.] Again, in electrokinetics the principle of minimum kinetic energy can be applied to establish the theory of induction currents. For instance, let there be a number of wires $C_1 \ldots C_n$ each forming a closed curve or circuit. Let electric currents be set up in these wires. If we denote by ϕ_1 the quantity of electricity that has passed in the positive direction through a section of the wire C_1 since a given epoch, the current in the wire C_1 at any instant will be represented by $\frac{d\phi}{dt}$ or $\dot{\phi}_1$.

With this notation the electrokinetic energy of the system at any instant is

$$T = \tfrac{1}{2} L_1 \dot{\phi}_1^2 + \tfrac{1}{2} L_2 \dot{\phi}_2^2 + \ldots$$
$$+ M_{12} \dot{\phi}_1 \dot{\phi}_2 + \ldots.$$

See Maxwell's *Electricity*, Vol. II. Art. 578.

In this expression the coefficient L_1 is

$$\iint \frac{\cos \epsilon}{r} ds ds',$$

where ds and ds' are two elements of the first circuit, $\frac{1}{r}$ the mean inverse distance between them*, ϵ the angle between their directions both taken the same way round the circuit, and the integration includes every pair of such elements. $L_2 \ldots L_n$ have corresponding values for the other circuits.

In like manner the coefficient M_{12} is

$$\iint \frac{\cos \epsilon}{r} ds_1 ds_2,$$

where ds_1 is an element of the first, and ds_2 of the second circuit, r and ϵ having the same meanings as before.

In the language of quaternions, if ρ_1 be the mean vector potential of the first circuit, ρ_2 that of the second, and so on,

* The wire having small finite thickness, let a be the distance from a point in a section of ds to a point in a section of ds'; then $\frac{1}{r}$ is the mean value of $\frac{1}{a}$ for all such pairs of points. In forming the mean vector potential for the circuit $\frac{1}{r}$ has the same signification.

$$L_1 = -\frac{1}{\dot{\phi}_1}\int S\rho_1\, ds_1,$$

$$M_{12} = -\frac{1}{\dot{\phi}_1}\int S\rho_1\, ds_2 = -\frac{1}{\dot{\phi}_2}\int S\rho_2\, ds_1,$$

&c. = &c.

Now, there being initially no currents in the wires, let a current $\dot{\phi}_1$ be suddenly generated in the wire C_1 by an electromotive force applied to that wire, all the wires remaining at rest in space. It is then observed that currents make their appearance simultaneously in the other wires. These are called *induced currents.*

Their values at the instant of the current $\dot{\phi}_1$ being created, that is before they sensibly decay by the resistance of the wires, are determined by the condition that the electrokinetic energy of the system is to be a minimum consistently with the existence of the current $\dot{\phi}_1$ in the wire C_1. That is by the equations

$$\frac{dT}{d\dot{\phi}_2} = 0, \ldots\ldots \frac{dT}{d\dot{\phi}_n} = 0,$$

in all as many equations as there are induction currents to be determined.

To take for simplicity the case of two circuits, if the current $\dot{\phi}_1$ be suddenly generated in C_1 by an electromotive force applied to C_1, then in order to determine the current $\dot{\phi}_2$ induced in C_2, we shall have the equation

$$\frac{dT}{d\dot{\phi}_2} = 0 \text{ or } M\dot{\phi}_1 + L_2\dot{\phi}_2 = 0;$$

that is,

$$\dot{\phi}_2 = -\frac{M}{L_2}\dot{\phi}_1.$$

Now L_2 is a necessarily positive quantity. Therefore $\dot{\phi}_2$ is in the same direction round the circuit as $\dot{\phi}_1$, or in the opposite direction, according as M is negative or positive.

If for instance both wires be circular and in parallel planes, and so placed that the projection of the first on the plane of the second lies outside the second, M will be negative, and $\dot{\phi}_2$ therefore in the same direction as $\dot{\phi}_1$. That is, both currents viewed from above may be in the same direction as the motion of the hands of a watch. It follows that if we compare those portions of the two wires which are nearest to each other, the current in C_2 will be in the opposite direction in space to that in C_1. If one be from south to north, the other will be from north to south. This agrees with the observed phenomena. See Maxwell, Art. 530.

If the current $\dot{\phi}_1$ be generated gradually, the rate at which the current $\dot{\phi}_2$ is destroyed by the resistance of the wire will generally bear a finite ratio to the rate at which it is generated by induction. But if the resistance be very small, we shall have the equation

$$\frac{d\dot{\phi}_2}{dt} = -\frac{M}{L_2}\frac{d\dot{\phi}_1}{dt}$$

to express the rate of variation of $\dot{\phi}_2$ in terms of that of $\dot{\phi}_1$. This is called *induction by variation of the primary circuit.*

Next let us consider the case of induction by *variation of the position of the circuits.* The currents in the two circuits being $\dot{\phi}_1$ and $\dot{\phi}_2$, let the second wire without change of its own shape be made to change its position in space relatively to the first wire, so that M will vary, L_1 and L_2 remaining constant. Further, let us suppose that no external electromotive force acts on either circuit, and that the resistance of the wires may be neglected. In that case we have by Lagrange's equations

$$\frac{d}{dt}\frac{dT}{d\dot{\phi}_1} - \frac{dT}{d\phi_1} = 0, \qquad \frac{d}{dt}\frac{dT}{d\dot{\phi}_2} - \frac{dT}{d\phi_2} = 0;$$

or since T is evidently independent of ϕ_1 and ϕ_2,

$$\frac{d}{dt}\frac{dT}{d\dot{\phi}_1} = 0, \qquad \frac{d}{dt}\frac{dT}{d\dot{\phi}_2} = 0;$$

that is,

$$L_1\frac{d\dot{\phi}_1}{dt} + M\frac{d\dot{\phi}_2}{dt} + \dot{\phi}_2\frac{dM}{dt} = 0,$$

$$M\frac{d\dot{\phi}_1}{dt} + L_2\frac{d\dot{\phi}_2}{dt} + \dot{\phi}_1\frac{dM}{dt} = 0;$$

whence

$$\frac{d\dot{\phi}_1}{dt} = \frac{dM}{dt}\cdot\frac{M\dot{\phi}_1 - L_2\dot{\phi}_2}{L_1L_2 - M^2},$$

$$\frac{d\dot{\phi}_2}{dt} = \frac{dM}{dt}\cdot\frac{M\dot{\phi}_2 - L_1\dot{\phi}_1}{L_1L_2 - M^2}.$$

If q be one of the coordinates defining the position of the second wire relatively to the first, we shall have evidently for any displacement denoted by ∂q

$$\frac{d\dot{\phi}_1}{dq} = \frac{dM}{dq} \cdot \frac{M\dot{\phi}_1 - L_2\dot{\phi}_2}{L_1 L_2 - M^2},$$

$$\frac{d\dot{\phi}_2}{dq} = \frac{dM}{dq} \cdot \frac{M\dot{\phi}_2 - L_1\dot{\phi}_1}{L_1 L_2 - M^2}.$$

Now the electromagnetic force tending to increase q is (Maxwell, vol. II. Art. 583)

$$\chi = \frac{dM}{dq} \dot{\phi}_1 \dot{\phi}_2.$$

The variation of χ due to the variation of the currents by induction is found by differentiating this expression regarding $\frac{dM}{dq}$ as a constant. That is,

$$\frac{d\chi}{dq} = \frac{dM}{dq} \cdot \left\{ \dot{\phi}_1 \frac{d\dot{\phi}_2}{dq} + \dot{\phi}_2 \frac{d\dot{\phi}_1}{dq} \right\}$$

$$= \left(\frac{dM}{dq}\right)^2 \cdot \frac{2M\dot{\phi}_1\dot{\phi}_2 - L_1\dot{\phi}_1^2 - L_2\dot{\phi}_2^2}{L_1 L_2 - M^2}.$$

Now L_1, L_2, and $L_1 L_2 - M^2$ are all necessarily positive, and therefore $\sqrt{L_1 L_2} - M$ is necessarily positive, and

$$2M\dot{\phi}_1\dot{\phi}_2 - L_1\dot{\phi}_1^2 - L_2\dot{\phi}_2^2$$

is necessarily negative. Hence the last equation shews that $\frac{d\chi}{dq}$ is necessarily negative. That is, the effect of displacing the second wire in any direction relatively to the first is to generate in the two wires induced currents which diminish the force tending to cause displacement in that direction. This agrees with the observed phenomena. See Maxwell, Art. 530.

CHAPTER III.

CHARACTERISTIC AND PRINCIPAL FUNCTIONS.

ARTICLE 36.] *Definition.* If T be the kinetic energy of any material system, and if A be equal to the definite integral

$$2\int_{t_0}^{t} T\,dt,$$

A is called *the Action* of the system from the time t_0 to the time t.

In any conservative system the Action between two given positions of the system may always be expressed in terms of the initial and final coordinates of the system and the total energy, and when thus expressed it satisfies the equations of which the following are types:*

$$\frac{dA}{dq} = p, \qquad \frac{dA}{dq_0} = -p_0, \text{ \&c.,} \qquad \frac{dA}{d\mathrm{E}} = t - t_0,$$

(q, p) being any coordinate and corresponding momentum in the final position, and (q_0, p_0) being the values of these magnitudes in the initial position, and E *being the total energy.*

For if the time t be reckoned from the beginning of the interval, the q's and the p's may by proper equations be expressed in terms of the q_0's, the initial values of the momenta, and the time t, and therefore T may be similarly expressed.

Also, if U be the force function, and E the total energy, we know that

$$T - U = \mathrm{E}.$$

By means of this equation and those last referred to we can eliminate t and express the initial momenta in terms of the q_0's, the q's, and E †.

* That is, a system in which the forces possess a force function.

† It is important to observe that the process in the text will give generally more than one *set* of initial momenta with which the system can pass from the given initial to the given final configuration. To each set corresponds a distinct type or value of A, and a distinct set of final momenta. The equations $\frac{dA}{dq} = p$, &c. hold for each type or value of A in relation to *the corresponding* momenta.

Hence T, and therefore A, may be similarly expressed in terms of the q_0's, the q's, and E, and therefore the first part of the proposition is proved.

Again, since

$$A = 2\int_{t_0}^{t} T dt,$$

$$\therefore \quad A = S\int_{t_0}^{t} mv^2 dt = S\int_{t_0}^{t} mv ds = \Sigma\int_{q_0}^{q} p dq \text{ by definition,}$$

where S denotes summation for all the elements of mass.

Let any possible variations be given to all the variables, then

$$\delta A = \Sigma\int_{q_0}^{q} p d\delta q + \Sigma\int_{q_0}^{q} \delta p dq.$$

Integrate the first term by parts and we get

$$\delta A = \Sigma\Big|p\delta q\Big|_{q_0}^{q} + \Sigma\int(\delta p dq - dp\delta q),$$

where

$$\Big|p\delta q\Big|_{q_0}^{q} = p\delta q - p_0\delta q_0;$$

$$\therefore \quad \delta A = \Sigma\Big|p\delta q\Big|_{q_0}^{q} + \Sigma\int_{t_0}^{t}(\dot{q}\delta p - \frac{dp}{dt}\delta q)dt.$$

But

$$\dot{q} = \frac{dT_p}{dp}, \quad \text{and} \quad \frac{dp}{dt} = F_q - \frac{dT_p}{dq};$$

$$\therefore \quad \delta A = \Sigma(p\delta q)_{q_0}^{q} + \Sigma\int_{t_0}^{t}(\frac{dT_p}{dp}\delta p + \frac{dT_p}{dq}\delta q - F_q\delta q)dt.$$

$$= \Sigma(p\delta q)_{q_0}^{q} + \int_{t_0}^{t}\delta T dt - \Sigma\int_{t_0}^{t} F_q\delta q dt.$$

But in this case $F_q = \dfrac{dU}{dq}$, the system being conservative;

$$\therefore \quad \Sigma\int_{t_0}^{t} F_q\delta q dt = \int_{t_0}^{t}\delta U dt;$$

$$\therefore \quad \delta A = \Sigma\Big|p\delta q\Big|_{q_0}^{q} + \int_{t_0}^{t}(\delta T - \delta U)dt$$

$$= \Sigma\Big|p\delta q\Big|_{q_0}^{q} + \int_{t_0}^{t}\delta \mathrm{E} dt$$

$$= \Sigma\Big|p\delta q\Big|_{q_0}^{q} + (t - t_0)\delta \mathrm{E};$$

$$\therefore \quad \frac{dA}{dq} = p, \qquad \frac{dA}{dq_0} = -p_0, \qquad \frac{dA}{d\mathrm{E}} = t - t_0.$$

When the Action A is thus expressed as a function of the initial and final coordinates and the energy it is written $f(q_0 \dots q \dots E)$, or more briefly f, and is called *the characteristic function.* It is clear that f satisfies a partial differential equation in the n variables $q_1, q_2 \dots q_n$ of the first order and the second degree, namely the equation which results from writing $\frac{df}{dq_1}, \frac{df}{dq_2}$, &c. for p_1, p_2, &c. in the equation

$$T_p = U + E.$$

37.] We next prove the converse of the last proposition:

If the partial differential equation in f be formed by writing $\frac{df}{dq_1}, \frac{df}{dq_2}$, &c. for p_1, p_2, &c. in the equation of conservation of energy of any conservative system, and if f be any solution of that equation, then an actual motion of the system may be determined by making $p_1 = \frac{df}{dq_1}$, $p_2 = \frac{df}{dq_2}$, &c., where p_1, p_2, &c. are the generalised components of momentum.

Since f satisfies the partial differential equation formed by substituting $\frac{df}{dq_1}, \frac{df}{dq_2}$, &c. for p_1, p_2, &c. in the equation

$$T_p = U + E,$$

therefore f must satisfy the n equations of the type

$$\frac{dT_p}{dq_r} + \frac{dT_p}{dp_1} \cdot \frac{dp_1}{dq_r} + \text{\&c.} + \frac{dT_p}{dp_n} \cdot \frac{dp_n}{dq_r} = \frac{dU}{dq_r} = F_r,$$

when $\frac{df}{dq_1}, \frac{df}{dq_2}$ &c. have been written for p_1, p_2, &c. in T_p.

But by Proposition (III)

$$\frac{dT_p}{dp_1} = \dot{q}_1, \quad \frac{dT_p}{dp_2} = \dot{q}_2, \text{ \&c.}$$

Therefore f satisfies the equation

$$\frac{dT_p}{dq_r} + \dot{q}_1 \frac{d^2 f}{dq_1 dq_r} + \dot{q}_2 \frac{d^2 f}{dq_2 dq_r} + \text{\&c.} = \frac{dU}{dq_r} = F_r;$$

$$\therefore \quad \frac{dT_p}{dq_r} + \left(\dot{q}_1 \frac{d}{dq_1} + \dot{q}_2 \frac{d}{dq_2} + \text{\&c.}\right) \frac{df}{dq_r} = F_r;$$

$$\therefore \quad \frac{dT_p}{dq_r} + \frac{d}{dt}\cdot\frac{df}{dq_r} = F_r;$$

$$\therefore \quad \frac{dT_p}{dq_r} + \frac{dp_r}{dt} = F_r,$$

if p_r be taken $= \frac{df}{dq_r}$; that is, the motion determined by making $p = \frac{df}{dq}$ satisfies Lagrange's equations, and is therefore a natural motion of the system.

38.] *If a complete primitive of the partial differential equation referred to in the last article be found in the form*

$$f(q_1 \dots q_n a_1 \dots a_{n-1}) + a_n,$$

where $a_1, a_2 \dots a_n$ are any arbitrary constants, then the integrals of the dynamical equations will be

$$\frac{df}{da_1} = \beta_1 \dots \frac{df}{da_{n-1}} = \beta_{n-1}, \quad \frac{df}{dE} = t + \epsilon,$$

where $\beta_1 \dots \beta_{n-1}\, \epsilon$ are n additional arbitrary constants.

For since f satisfies the partial differential equation

$$T_p = U + E, \dots \dots \dots \dots (1)$$

when p_1, p_2, &c. have been replaced by $\frac{df}{dq_1}, \frac{df}{dq_2}$, &c., it follows that f must satisfy the $n-1$ equations of the type

$$\frac{dT_p}{dp_1}\frac{dp_1}{da_1} \dots + \frac{dT_p}{dp_n}\frac{dp_n}{da_1} = 0,$$

with the additional equation

$$\frac{dT_p}{dp_1}\cdot\frac{dp_1}{dE} + \dots + \frac{dT_p}{dp_n}\frac{dp_n}{dE} = 1,$$

found by differentiating (1) with regard to $a_1, \dots a_{n-1}$ and E successively.

Now, with the substitutions referred to, $\frac{dT_p}{dp_1} = \dot{q}_1$, &c., these equations become

$$\left.\begin{array}{c} \dot{q}_1 \frac{d^2 f}{dq_1 da_1} + \&\text{c.} + \dot{q}_n \frac{d^2 f}{dq_n da_1} = 0 \\ \&\text{c.} \\ \dot{q}_1 \frac{d^2 f}{dq_1 dE} + \&\text{c.} + \dot{q}_n \frac{d^2 f}{dq_n dE} = 1 \end{array}\right\} \dots \dots (2)$$

from which $\dot{q}_1, \dot{q}_2, \dots \dot{q}_n$ can be determined.

But if we differentiate with regard to t the n equations

$$\frac{df}{da_1}=\beta_1 \dots \frac{df}{da_{n-1}}=\beta_{n-1}, \quad \frac{df}{d\mathrm{E}}=t+\mathrm{E},$$

we obtain precisely these last equations (2) to determine the magnitudes $\dot{q}_1, \dot{q}_2 \dots \dot{q}_n$.

For instance, differentiating $\frac{df}{da_1}=\beta_1$ with regard to t we obtain

$$\frac{d}{dt}\frac{df}{da_1}=(\dot{q}_1\frac{d}{dq_1}+\dots+\dot{q}_n\frac{d}{dq_n})\frac{df}{da_1}=0.$$

Whence the proposition is established.

39.] *Definition.* If A represent the Action in any conservative system where the time does not enter explicitly into the connecting equations, and if S be determined as a function of the initial and final coordinates and the time, by means of the equation

$$S=A-\mathrm{E}(t-t_0),$$

the function thus found is called the *Principal Function**.

If S be the principal function in any conservative system where the time t does not enter explicitly into the connecting equations, then

$$\frac{dS}{dq}=p, \quad \frac{dS}{dq_0}=-p_0, \quad \frac{dS}{dt}=-\mathrm{E},$$

where q_0 represents any one of the initial, and q any one of the final coordinates.

For since
$$S=A-\mathrm{E}t,$$

$$\therefore \quad \delta S=\delta A-\mathrm{E}\,\delta t-t\,\delta\mathrm{E};$$

and therefore if the final coordinates alone vary,

$$\Sigma\frac{dS}{dq}\delta q+\frac{dS}{dt}\delta t=\Sigma\frac{dA}{dq}\delta q+\frac{dA}{d\mathrm{E}}\cdot\delta\mathrm{E}-\mathrm{E}\,\delta t-t\,\delta\mathrm{E}.$$

But
$$\frac{dA}{dq}=p \quad\text{and}\quad \frac{dA}{dE}=t,$$

$$\therefore \quad \Sigma\frac{dS}{dq}\delta q+\frac{dS}{dt}\delta t=\Sigma p\,\delta q-\mathrm{E}\,\delta t;$$

$$\therefore \quad \frac{dS}{dq}=p, \quad \frac{dS}{dt}=-\mathrm{E};$$

* The time is so very generally reckoned from the beginning of the motion that unless the contrary be expressly mentioned it will be assumed that $t_0=0$.

and by varying the initial coordinates we obtain similarly,

$$\frac{dS}{dq_0} = -p_0.$$

40.] Since A is a function of the initial and final coordinates, represented typically by q_0 and q, and of E, the increase of A per unit of time as the system passes through the configuration q is clearly

$$\frac{dA}{dt} = \Sigma \dot{q} \frac{dA}{dq}.$$

We may conceive the system passing through the same configuration q with any other velocities $\dot{q}'$ and the same value of E. The increase of A, considered as a function of q_0 and q, per unit of time in this latter motion is

$$\Sigma \dot{q}' \frac{dA}{dq}.$$

Now, since the kinetic energy in the configuration q is the same for both motions and

$$p = \frac{dA}{dq},$$

it follows from Art. 10 that the increase of A per unit of time is greater in the actual motion than in the $\dot{q}'$ motion, and, given E, is a maximum in the actual motion. This is true for every material system. In the case of a free particle of mass m we have

$$p = m\dot{q},$$

q_1, q_2, q_3 being the rectangular coordinates of the particle, and the equation

$$p = \frac{dA}{dq}$$

in this case expresses the fact that the path of the particle is normal to the surfaces of equal action. By extending the meaning of the terms 'normal' and 'surface,' we might say generally that a motion in which p is proportional to $\frac{df}{dq}$ is normal to the surface $f =$ constant.

If t instead of E be invariable, similar statements apply to the principal function S.

41.] As an illustration of the formation of the principal and characteristic functions, let us consider the case of a projectile of mass unity. Let the point of projection be the origin of coordinates, x and y the horizontal and vertical coordinates of the projectile. Let the initial and terminal horizontal and vertical velocities be u_0, v_0, u, v. Let the time t be measured from the instant of projection. Let the potential at the point of projection be zero. Here we have

$$u^2+v^2+2gy = u_0^2+v_0^2 = 2\text{E};$$

$$u = u_0, \qquad v = v_0 - gt;$$

$$x = u_0 t, \qquad y = v_0 t - \frac{gt^2}{2} = vt + \frac{gt^2}{2};$$

$$x^2+\left(y+\frac{gt^2}{2}\right)^2 = (u_0^2+v_0^2)t^2 = 2\text{E}t^2;$$

$$(\alpha) \qquad \therefore \quad \text{E} = \frac{1}{2t^2}\left\{x^2+\left(y+\frac{gt^2}{2}\right)^2\right\};$$

$$(\beta) \qquad t^2 = \frac{2}{g^2}\{2\text{E}-gy \pm \sqrt{4\text{E}^2-4\text{E}gy-g^2x^2}\},$$

$$A = \int_0^t (u^2+v^2)\,dt = u_0^2 t - \frac{1}{3g}\{(v_0-gt)^3-v_0^3\}$$

$$= \frac{x^2}{t} + \frac{1}{3g}\left\{\left(\frac{y}{t}+\frac{gt}{2}\right)^3 - \left(\frac{y}{t}-\frac{gt}{2}\right)^3\right\};$$

$$(\gamma) \qquad = \frac{x^2+y^2}{t} + \frac{g^2}{12}t^3;$$

$$(\delta) \qquad S = A - \text{E}t = \frac{x^2+y^2}{2t} - \frac{g^2t^3}{24} - \frac{gyt}{2}.$$

The expression for S on the right-hand side of δ is the Principal Function, and on being differentiated with regard to x, y, and t respectively, attention being paid to (α) we shall obtain the quantities u, v, and $-$E.

The expression for A on the right-hand side of (γ) is not in its present form the Characteristic Function, but we may obtain that function by substituting in (γ) the value of t obtained from (β).

Thus let A in (γ) be differentiated with regard to y, x and E being constant, and we get

$$\frac{dA}{dy} = \frac{2y}{t} + \frac{dt}{dy}\left\{\frac{g^2t^2}{4} - \frac{x^2+y^2}{t^2}\right\}$$

$$= \frac{2y}{t} + \frac{dt}{dy}\left\{\frac{g^2t^2}{2} + gy - 2\text{E}\right\} \text{ by } (\alpha);$$

$$= \frac{2y}{t} + \frac{dt}{dy}(v_0 gt - 2\text{E}).$$

Also from (β)

$$t\frac{dt}{dy} = -\frac{1}{g} \mp \frac{1}{g}\frac{2\text{E}}{\sqrt{4\text{E}^2 - 4\text{E}gy - g^2x^2}},$$

the + sign being used if the − sign be used in the expression for t^2, and *vice versa.* Also from (β),

$$\pm\sqrt{4\text{E}^2 - 4\text{E}gy - g^2x^2} = \frac{g^2t^2}{2} + gy - 2\text{E},$$

$$= v_0 gt - 2\text{E}.$$

Hence we obtain

$$\frac{dt}{dy}(v_0gt - 2\text{E}) = -\frac{v_0gt - 2\text{E}}{gt} - \frac{2\text{E}}{gt},$$

$$= -v_0;$$

and $$\frac{dA}{dy} = \frac{2y}{t} - v_0 = v.$$

By the same process we may obtain

$$\frac{dA}{dx} = u.$$

If in the above formulae we were to write $x-x_0$, $y-y_0$ for x and y, taking x_0, y_0 for the initial coordinates, we might obtain by the same method

$$\frac{dA}{dx_0} = -u_0, \quad \frac{dA}{dy_0} = -v_0.$$

42.] As another example of the formation and properties of the Characteristic and Principal Functions we may take the case of the elliptic orbit under a central force μr.

The equations of motion are in this case

$$\frac{d^2x}{dt^2} + \mu x = 0; \quad \ldots \quad (1)$$

$$\frac{d^2y}{dt^2} + \mu y = 0. \quad \ldots \quad (2)$$

The integrals are

$$x = a\cos\sqrt{\mu}t + b\sin\sqrt{\mu}t, \qquad y = a'\cos\sqrt{\mu}t + b'\sin\sqrt{\mu}t;$$

$$\frac{dx}{dt} = \sqrt{\mu}\{b\cos\sqrt{\mu}t - a\sin\sqrt{\mu}t\},$$

$$\frac{dy}{dt} = \sqrt{\mu}\{b'\cos\sqrt{\mu}t - a'\sin\sqrt{\mu}t\}.$$

Whence we easily get (remembering that the force function U is

$$-\frac{\mu}{2}(x^2+y^2)),$$

$$2\mathrm{E} = \mu(a^2+b^2+a'^2+b'^2),$$

$$\mathrm{E}t = \frac{\mu}{2}(a^2+b^2+a'^2+b'^2)t;$$

$$A \text{ or } \int_0^t (\frac{dx}{dt})^2 dt + \int_0^t (\frac{dy}{dt})^2 dt$$

$$= \mu[\frac{a^2+b^2}{2}t + \frac{1}{2\sqrt{\mu}}\{\frac{b^2-a^2}{2}\sin 2\sqrt{\mu}t - ab(1-\cos 2\sqrt{\mu}t)\}]$$

$$+\mu[\frac{a'^2+b'^2}{2}t + \frac{1}{2\sqrt{\mu}}\{\frac{b'^2-a'^2}{2}\sin 2\sqrt{\mu}t - a'b'(1-\cos 2\sqrt{\mu}t)\}];$$

and

$$S = A - \mathrm{E}t = A - \frac{\mu}{2}(a^2+b^2+a'^2+b'^2)t$$

$$= \frac{\sqrt{\mu}}{2}\{\frac{b^2+b'^2-a^2-a'^2}{2}\sin 2\sqrt{\mu}t - (ab+a'b')(1-\cos 2\sqrt{\mu}t)\}.$$

But if x_0, y_0 be the initial coordinates, we have

$$x_0 = a, \qquad y_0 = a';$$

$$b = \frac{x - x_0\cos\sqrt{\mu}t}{\sin\sqrt{\mu}t}, \qquad b' = \frac{y - y_0\cos\sqrt{\mu}t}{\sin\sqrt{\mu}t};$$

whence by substitution S is easily reduced to

$$\frac{\sqrt{\mu}}{2}\{(x^2+y^2+x_0^2+y_0^2)\cot\sqrt{\mu}t - 2(xx_0+yy_0)\operatorname{cosec}\sqrt{\mu}t\}.$$

And thus the principal function is found. It will be seen also that S satisfies the two differential equations

$$\frac{dS}{dt} + \frac{1}{2}\left\{(\frac{dS}{dx})^2 + (\frac{dS}{dy})^2\right\} = -\frac{\mu}{2}(x^2+y^2),$$

$$\frac{dS}{dt} + \frac{1}{2}\left\{(\frac{dS}{dx_0})^2 + (\frac{dS}{dy_0})^2\right\} = -\frac{\mu}{2}(x_0^2+y_0^2).$$

To find A we must first of all determine t as a function of x_0, y_0, x, y, and E, and then eliminate t from the expression for S.

Now

$$\frac{2\mathrm{E}}{\mu} = a^2 + a'^2 + b^2 + b'^2$$

$$= x_0^2 + y_0^2 + \frac{(x - x_0 \cos\sqrt{\mu}\,t)^2 + (y - y_0 \cos\sqrt{\mu}\,t)^2}{\sin^2\sqrt{\mu}\,t},$$

$$= \frac{x^2 + y^2 + x_0^2 + y_0^2 - 2(xx_0 + yy_0)\cos\sqrt{\mu}\,t}{\sin^2\sqrt{\mu}\,t};$$

$$\therefore \quad \cos^2\sqrt{\mu}\,t - \frac{\mu}{\mathrm{E}}(xx_0 + yy_0)\cos\sqrt{\mu}\,t = 1 - \frac{\mu}{2\mathrm{E}}(x^2 + y^2 + x_0^2 + y_0^2);$$

$$\therefore \quad \cos\sqrt{\mu}\,t = \frac{\mu}{2\mathrm{E}}(xx_0 + yy_0)$$

$$\pm\sqrt{1 - \frac{\mu}{2\mathrm{E}}(x^2 + y^2 + x_0^2 + y_0^2) + \frac{\mu^2}{4\mathrm{E}^2}(xx_0 + yy_0)^2}.$$

If the value of t thus found be substituted in S and Et be added, we obtain the characteristic function A, and it will be found that A satisfies the equations

$$\frac{dA}{dx} = \frac{dx}{dt}, \quad \frac{dA}{dx_0} = -\frac{dx}{dt_0}, \quad \frac{dA}{dy} = \frac{dy}{dt}, \quad \frac{dA}{dy_0} = -\frac{dy}{dt_0};$$

and the partial differential equations

$$\frac{1}{2}\left\{\left(\frac{dA}{dx}\right)^2 + \left(\frac{dA}{dy}\right)^2\right\} = -\frac{\mu}{2}(x^2 + y^2) + \mathrm{E},$$

$$\frac{1}{2}\left\{\left(\frac{dA}{dx_0}\right)^2 + \left(\frac{dA}{dy_0}\right)^2\right\} = -\frac{\mu}{2}(x_0^2 + y_0^2) + \mathrm{E}.$$

CHAPTER IV.

STATIONARY AND LEAST ACTION.

ARTICLE 43.] Let a material system be in motion under the action of any conservative forces, and in the interval between the times t_0 and t let it pass from any given configuration to any other.

Let A be the action between these two configurations so that

$$A = \Sigma \int_{t_0}^{t} m\, v^2 dt,$$

and let T be the kinetic energy, U the force function, and E the total energy at any instant during the motion.

Let the motion of the system be ideally varied, so that while the initial and final configurations remain the same as before the system shall pass from one to the other through a series of configurations always indefinitely near to some configuration in the actual motion, and also so that the equation

$$T - U = \mathrm{E}$$

remains true for the same value of E throughout the varied motion. Such a varied motion is ideally possible, but can in general only be effected actually by the introduction of additional constraints from without. Then, in such a case, the small variation δA in the value of the action in passing from the original unconstrained to the varied constrained motion is always zero. This is the principle of *Stationary Action.*

For
$$A = \Sigma \int_{t_0}^{t} m v^2 dt = \Sigma \int_{q_0}^{q} p\, \delta q;$$

therefore, as in Art. 36,

$$\delta A = \Sigma \left(p \delta q\right)_{q_0}^{q} + \Sigma \int_{t_0}^{t} \left(\dot{q} \delta p - \frac{dp}{dt} \delta q\right) dt$$

for any small variations whatever δp and δq.

And this equation reduces, as in the article mentioned, to

$$\delta A = \Sigma (p\delta q)_{q_0}^{q} + \int_{t_0}^{t} (\delta T - \delta U)\, dt.$$

But the initial and final coordinates q_0 and q remain unaltered, as also does $T-U$ by our hypothesis,

$$\therefore \quad \{\Sigma p\,\delta q\}_{q_0}^{q} = 0 \quad \text{and} \quad \delta T - \delta U = 0;$$

$$\therefore \quad \delta A = 0.$$

44.] Exactly in the same way it may be shewn that if the time be the same in the two courses, but E vary, then when the initial and final coordinates remain unaltered, $\delta S = 0$.

For $$\delta S = \delta A - (t-t_0)\,\delta \mathrm{E} = \delta A - (t-t_0)\,\{\delta T - \delta U\},$$

$$= \Sigma (p\delta q)_{q_0}^{q} + \int_{t_0}^{t} (\delta T - \delta U)\, dt - (t-t_0)(\delta T - \delta U),$$

$$= 0.$$

45.] In the above expressions δA and δS respectively include the first powers only of small variations according to the ordinary notation of the Differential Calculus and the Calculus of Variations; and the Principle of Stationary Action just proved shews that the difference between the actions in the original and varied motion is to the first approximation zero; A therefore satisfies the first condition of being a minimum.

We proceed now to investigate the sign of this difference when higher orders of the variations are considered, and the final result will be to shew that when certain conditions are fulfilled the Action will be a true minimum, and that when these conditions are not fulfilled, no general rule can be asserted concerning it.

Before treating the general question we will consider the simple case of the projectile.

For the sake of brevity we will again suppose the origin to be the point of projection; then as before

$$u_0^2 + v_0^2 \quad \text{or} \quad V^2 = 2\mathrm{E},$$

where V is the velocity of projection. Let also a be the angle of projection so that

$$u_0 = V\cos a, \quad v_0 = V\sin a.$$

Then, if t be the time in passing to the point x, y,

$$t^2 = \frac{2}{g^2}(V^2 - gy) \pm \frac{2}{g^2}\sqrt{V^4 - 2V^2gy - g^2x^2}; \quad \ldots \ldots \quad \text{(I)}$$

and if a be the angle of projection, in order that the projectile may pass through x, y,

$$\tan a = \frac{V^2 \pm \sqrt{V^4 - 2V^2gy - g^2x^2}}{gx}. \quad \ldots \ldots \ldots \quad \text{(II)}$$

From (II) we obtain

$$V \sin a \quad \text{and} \quad V \cos a,$$

or the initial values of the momenta in terms of V, x, y.

Again, as we found above,

$$A = \frac{x^2 + y^2}{t} + \frac{g^2}{12}t^3, \quad \ldots \ldots \ldots \ldots \ldots \quad \text{(III)}$$

and if we substitute for t in this expression the value obtained in (I) we obtain the characteristic function f, or the expression for A in terms of x, y, and V^2 (*i.e.* 2E).

From (I) we see that

$$t = \pm \sqrt{\frac{2}{g^2}(V^2 - gy) \pm \frac{2}{g^2}\sqrt{V^4 - 2V^2gy - g^2x^2}}.$$

And as a negative value of t has no meaning bearing upon our question, we shall reject it and take

$$t = \sqrt{\frac{2}{g^2}(V^2 - gy) \pm \sqrt{V^4 - 2V^2gy - g^2x^2}}.$$

Similarly, in determining $V \sin a$ and $V \cos a$, we shall take for $\sin a$ and $\cos a$, in terms of $\tan a$, the values

$$+\frac{\tan a}{\sqrt{1 + \tan^2 a}} \quad \text{and} \quad +\frac{1}{\sqrt{1 + \tan^2 a}} \quad \text{respectively.}$$

Whence it appears that when the time and the two initial momenta are determined in terms of x, y, and V^2, each of these quantities will be expressed by two distinct functions of x, y, and V^2, owing to the double sign of the radical

$$\sqrt{V^4 - 2V^2gy - g^2x^2};$$

and the same may be said of A, since by (III) there corresponds a value of A to each value of t.

Hence we learn that if a particle be projected from O with a given velocity, and be required to pass through the point P whose coordinates referred to O are x and y, the necessary horizontal and vertical momenta at O and the action from O to P will be given in each case by either of two distinct functions of x, y, and V^2, so that *in general* there are two distinct courses from O to P, viz. OC_1P and OC_2P, having different times of passage, different initial momenta, and different values of the Action.

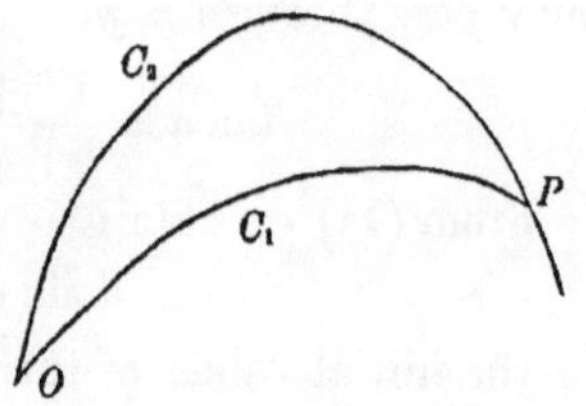

Fig. 2.

If however the point P be so situated that

$$V^4 - 2V^2gy - g^2x^2 = 0,$$

then the radical vanishes, the two functions mentioned above coincide in value in the expressions for momenta, time, and Action, and the two courses from O to P become coincident in all respects.

The locus of P thus determined is clearly the parabola ACC', touching the common directrix of all the parabolas at the point A, vertically above the point of projection O, and having its focus at O.

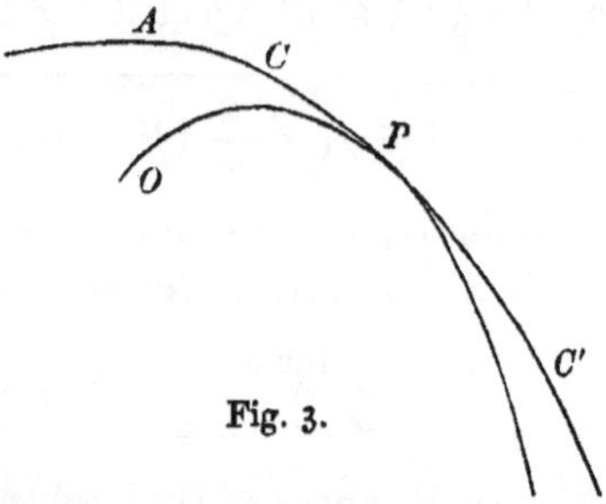

Fig. 3.

If we find the envelope of the curves

$$y = x\tan a - \frac{gx^2}{2V^2\cos^2 a}$$

for the variable parameter a, we obtain the locus ACC', whence it appears that the path described by each one of the bodies projected from O with the velocity V touches the parabola ACC'.

If in the equation $V^4 - 2V^2gy - g^2x^2 = 0$ we substitute for x and y their values $u_0 t$ and $v_0 t - \frac{gt^2}{2}$ respectively, we obtain

$V^2 - gv_0 t = 0$, or $t = \frac{V^2}{gv_0}$, giving the time from O to the point of contact with the envelope, which is then positive if v_0 be positive.

The following conclusions may now be drawn:

(1) If a point be taken outside the parabola ACC' it cannot be reached by a body projected from O with the given velocity V, because for such points

$$V^4 - 2V^2 gy - g^2 x^2$$

being negative, the formulae above obtained give imaginary values for t, a, and A.

(2) If a point be taken within the parabola ACC', it can be reached by a body projected from O with the given velocity V in two different directions, giving rise to two distinct courses in which the initial momenta, the times of flight, and the Action have different values, one of these courses (OCP) reaching P after touching the envelope, and the other (OPQ) either touching the envelope after passing through P, or not at all, according as the direction of projection in it is above or below the horizontal. If the direction of projection from O be below the horizontal, the course touches the envelope at a point to the left of O, that is at a point for which t is negative.

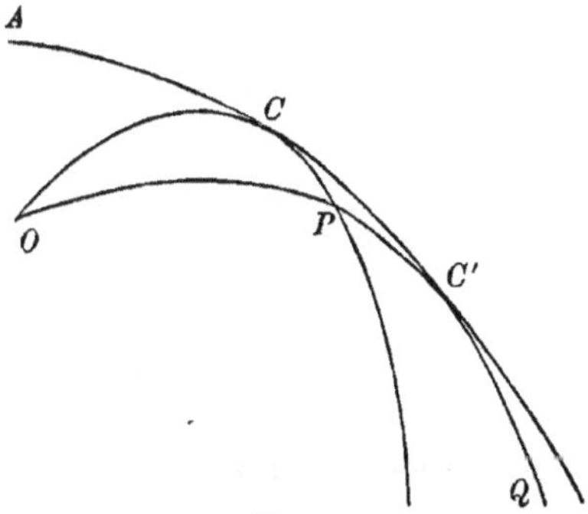

Fig. 4.

(3) If P and Q (Fig. 4) be two points taken on one of the projectile paths, P before and Q beyond the point of contact (C') of that path with the envelope, then the initial momenta, time, and Action expressed in terms of the coordinates of P and Q respectively, will be each of them given by taking in the one

case the negative and in the other the positive value of the radical. This is made clear from the expression in (II)

$$\tan a = \frac{V^2}{gx} \pm \frac{\sqrt{V^4 - 2V^2gy - g^2x^2}}{gx}.$$

Whence it follows that

$$\tan a = \frac{V^2}{gx_1},$$

where x_1 is the abscissa of C'.

But the abscissa at P is less than x_1, and that at Q is greater than x_1, and V is constant.

It follows therefore that in the expression for $\tan a$ in terms of the coordinates of P or of Q, when the coordinates of P are substituted the numerator must be less than V^2, and when the coordinates of Q are substituted the numerator must be greater than V^2, i.e. the negative sign of the radical must be taken between O and C', and the positive sign must be taken beyond C'.

It thus appears that neither the time, the initial momenta, nor the Action is expressed by one and the same function of the coordinates throughout the whole course. The function in each case expressing these quantities changes its form or type at the point of contact with the envelope.

(4) Of the two courses from O to P, that one which reaches P after touching the envelope has the greater Action. For, let t_1, t_2 be the times, A_1, A_2 the Actions, in the two courses; then from (III)

$$A_1 - A_2 = (t_1 - t_2)\left\{\frac{g^2}{12}(t_1^2 + t_2^2 + t_1 t_2) - \frac{x^2+y^2}{t_1 t_2}\right\}.$$

Also from (β), Art. 41, above

$$t_1^2 + t_2^2 = \frac{4}{g^2}(V^2 - gy),$$

$$t_1 t_2 = \frac{2}{g^2}\{(V^2-gy)^2 - (V^4 - 2V^2gy - g^2x^2)\}^{\frac{1}{2}},$$

$$= \frac{2}{g}\sqrt{x^2+y^2}.$$

Therefore

$$A_1 - A_2 = (t_1 - t_2)\left\{\frac{V^2 - gy - g\sqrt{x^2+y^2}}{3}\right\};$$

but since

$$V^4 - 2V^2gy - g^2x^2 \text{ is positive,}$$

$$V^2 - gy > g\sqrt{x^2+y^2}.$$

Therefore $A_1 - A_2$ has the same sign as $t_1 - t_2$; and since the course which reaches P after touching the envelope has the greater time, it has the greater Action. It follows that if A_1, A_2 be the two functions expressing the Action from O to P, $A_1 - A_2$ is for real values of A_1 and A_2 essentially negative if the course having Action A_2 reach P after touching the envelope.

(5) The curves of equal Action are as above proved normal to the courses, and therefore when they meet the enveloping parabola must be at right angles to it. It follows that the curve A = constant has two branches, forming a cusp when it meets the envelope, and one branch, the upper, intersects the courses orthogonally after they touch the envelope.

46.] *Definition.*—The point C to which the two courses from O coincide is called a *kinetic focus conjugate* to O. Evidently it is the point in which the courses touch the envelope.

We may now shew that in the case of the projectile the Action from O to P in the natural course is less than it would be in any infinitely near constrained course provided P lie between O and C, the kinetic focus conjugate to O in the natural course OPC, where the Action changes type.

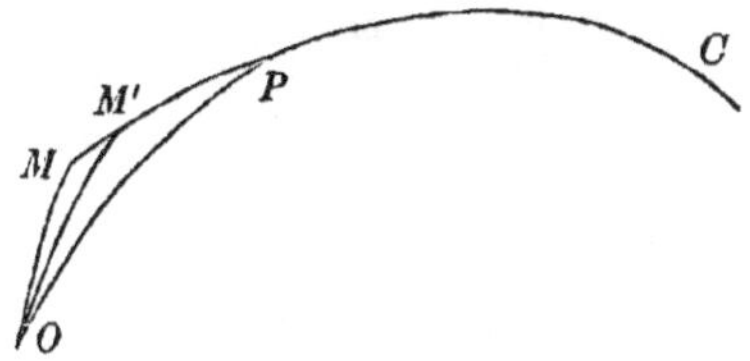

Fig. 5.

Let P lie between O and C, and let OM be a natural course, very near to OP, such that M may be reached by a projectile starting from O with the same energy as that in OP, and with initial momenta the same functions of the coordinates of M, as those in OP are of the coordinates of P, *i.e.* with the use in each case of the *negative* value of the radical spoken of in the discussion of the last article. At M suppose a fresh infinitely small impulse applied to the particle so as not to change its total energy and therefore not to change its velocity at that point but to cause it to reach P by another projectile trajectory MP, infinitely near OP.

It is clear that any point M' infinitely near M in the course MP may be reached by a trajectory OM' starting with the same total energy, therefore with the same velocity, as in OP and with the same type or sign of the radical.

Let p and $\dot{q}$ be either generalised momentum and corresponding velocity at M in OM, and let p', $\dot{q}'$ be corresponding quantities at M in the course MM'. Let A be the Action in the course OM from O to M, and $A+\delta A$ the Action in the course OM' from O to M'.

Then, from the general proposition

$$\frac{dA}{dq} = p$$

we have

$$\delta A = \Sigma p\,\delta q.$$

Also $\delta q = \dot{q}'\delta t$ if δt be the time from M to M' in the course MM',

$$\therefore \quad \delta A = \Sigma p\,\dot{q}'\,\delta t.$$

Similarly $\delta' A$, the Action from M to M', $= \Sigma p'\dot{q}'\,\delta t$.

Therefore

$$\text{Action } OM' <> \text{Action } OM + \text{Action } MM'$$

as

$$\Sigma p\dot{q}' <> \Sigma p'\dot{q}'.$$

But since E and U are respectively the same at M in the courses OM and MM', it follows that T must be the same in both courses, i.e. $\Sigma p\dot{q} = \Sigma p'\dot{q}'$.

Therefore, by Proposition ~~IX~~, VIII

$$\Sigma p\dot{q}' < \Sigma p'\dot{q}'.$$

Therefore

$$\text{Action } OM' < \text{Action } OM + \text{Action } MM'.$$

Similarly, if M'' be a point in $M'P$ very near M',

$$\text{Action } OM'' < \text{Action } OM' + \text{Action } M'M''.$$

And so on,

$$\text{Action } OP < \text{Action } OM + \text{Action } MP.$$

It is clear that every constrained course from O to P may be broken up into a number of natural courses, and that by the continued application of this proposition we shall always have—

Action OP less than sum of the Actions in the broken course, or Action OP less than the Action in the constrained course.

If now the point were taken on the course OPC, beyond C, as at Q (Fig. 6), we cannot make use of the previous reasoning, because the use of the equation $\frac{dA}{dq} = p$ implies that the same sign of the radical is taken throughout in the expression for A. If then the Action in OM have the negative sign of the radical, M lying very near some point in the course between O and C, and if, as before, we draw a series of natural courses from O to points in MQ, the Action in all such courses must, in order that the proposition may be applied, have the same sign of the radical, that is the negative sign. But the Action from O to Q in OCQ has, as we have seen, the positive sign. If then the constrained course MQ be so drawn as not to meet the envelope, the continued application of the proposition would result in proving that Action $(OM+MQ)$ is greater, not than Action OCQ, but than Action $OC'Q$, the Action in the other, and as we have proved the shorter, course from O to Q.

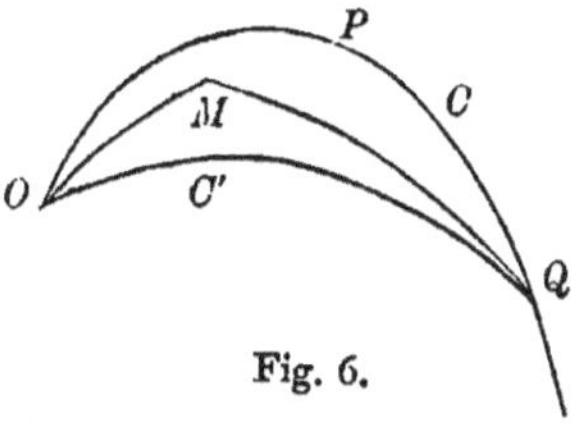

Fig. 6.

But if the point M in Fig. 6 were so taken as that the Action in OM in a course infinitely near to OCQ should have the positive sign of the radical, that is if M were taken beyond the kinetic focus C, the proposition might be applied to shew that Action $(OM+MQ)$ is greater than Action OCQ.

As we have already shewn that a natural course exists from O to Q having less Action than OCQ, it is easily seen that some constrained course exists having less Action. For instance, let Q' be very near C and beyond it (Fig. 7), $OC'Q'$ the course of less Action from O to Q', M a point in $OC'Q'$ very near Q'. Let the ~~system~~ particle receive at M any small impulse not altering its kinetic energy, so as to cause it to describe a new trajectory infinitely near $OCQ'Q$ and intersecting $OCQ'Q$ in Q.

Then by the above method it may be shewn that

$$\text{Action }(MQ' + Q'Q) > \text{Action } MQ,$$

$$\therefore \quad \text{Action }(OQ' + Q'Q) > \text{Action }(OM + MQ),$$

$$\therefore \quad \textit{à fortiori} \text{ Action } OCQ > \text{Action }(OM + MQ);$$

and $OM + MQ$ is a constrained course infinitely near OCQ.

It appears therefore that in the case of the projectile the Action from O to any point in the course is a true minimum, so long as it is represented by *the same function* of the initial and final coordinates, and ceases to be a minimum when the function changes.

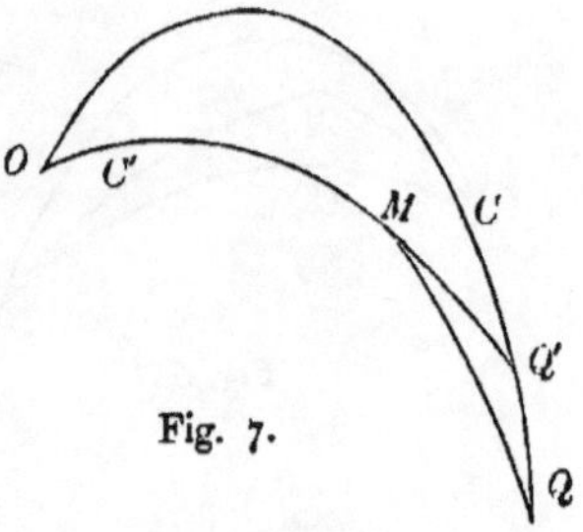

Fig. 7.

47.] An additional illustration of this subject may be derived from the second of the two examples investigated above, namely, that of the ellipse under the action of a force varying as the distance.

In that case we have seen that

$$S = \frac{\sqrt{\mu}}{2}\left\{(x^2+y^2+x_0^2+y_0^2)\cot\sqrt{\mu}t - 2\,(x x_0 + y y_0)\operatorname{cosec}\sqrt{\mu}t\right\}, \quad \text{(I)}$$

and that

$$A = S + \frac{\mu}{2}\,\frac{x^2+y^2+x_0^2+y_0^2-2\,(xx_0+yy_0)\cos\sqrt{\mu}t}{\sin^2\sqrt{\mu}t}\,t; \quad .. \quad \text{(II)}$$

where in (II) we must substitute for t the value given by the equation

$$\cos\sqrt{\mu}t = \frac{\mu}{2E}(xx_0+yy_0)$$

$$\pm\sqrt{1-\frac{\mu}{2\text{E}}(x^2+y^2+x_0^2+y_0^2)+\frac{\mu^2}{4\text{E}^2}(xx_0+yy_0)^2}, \quad .. \quad \text{(III)}$$

E being the total energy.

The value of A thus obtained will be the total Action from any given initial to any final configuration in terms of the coordinates of the two configurations. The differential coefficients of A with regard to x_0 and y_0 respectively, after the required substitution for t, give us the requisite initial momenta to enable the particle to pass from the initial to the final configurations in terms of the coordinates of those configurations.

It appears from (III) that there are two distinct values of $\cos\sqrt{\mu}t$ in terms of the coordinates of the two configurations, and therefore two distinct elliptic orbits, by either of which the particle may move from O, the point of projection, to P with given initial kinetic energy. Again, in either orbit the motion may be either direct or retrograde. In each ellipse the value of $\cos\sqrt{\mu}t$ is the same for the direct as for the retrograde motion, but $\sqrt{\mu}t$ is represented in the one case (which we may call the direct motion) by θ, and in the other by $2\pi-\theta$, θ being a positive angle less than π, and having a single value for every point in the orbit. Regarding only the two direct motions from O to P, we shall obtain two distinct values of each initial momentum, and of the Action, *i.e.* two courses expressible by two distinct functions f_1 and f_2 of the initial and final coordinates*. Similarly, if we regard the two retrograde motions, we shall obtain two other distinct courses from O to P, expressible by the functions f_3 and f_4 of the initial and final coordinates.

The two ellipses will coincide when the values of x, y cause the radical in (III) to vanish, and in that case the two direct courses coincide, and likewise the two retrograde courses.

This locus of x, y is clearly an ellipse, and if the starting-point be taken at a distance c, from the origin on the axis of x so that $a = c$, $a' = 0$, and if the initial velocity be V so that

$$2\,\mathrm{E} = V^2+\mu c^2,$$

the locus easily reduces to

$$\frac{x^2}{V^2+\mu c^2}+\frac{y^2}{V^2}=\frac{1}{\mu}.$$

* We here neglect all the other values arising from the expression $t=\cos^{-1}m$ when m is known, because these only correspond to the return of the particle to the point $\pm x$, $\pm y$ after successive revolutions.

If we express the integrals of our equations of motion in the form

$$x = a\cos(\sqrt{\mu}\,t + a),$$
$$y = b'\sin(\sqrt{\mu}\,t),$$

we find that the particle describes an ellipse whose equation is

$$\frac{x^2}{a^2\cos^2 a} + \frac{2\sin a\,xy}{b'\,a\cos^2 a} + \frac{y^2}{b'^2\cos^2 a} = 1,$$

where $$V^2 = \mu(a^2\sin^2 a + b'^2);$$

and the coordinates x_0, y_0 of the point of projection are $a\cos a$ and 0; putting c for x_0 the equations become

$$\frac{x^2}{c^2} + \frac{2\tan a\,xy}{b'c} + \frac{y^2}{b'^2\cos^2 a} = 1; \quad \ldots\ldots\ldots\ldots \quad \text{(A)}$$

$$V^2 = \mu(c^2\tan^2 a + b'^2). \quad \ldots\ldots\ldots\ldots\ldots\ldots \quad \text{(B)}$$

If we investigate the envelope of (A), with b' and a variable parameters subject to the condition (B), we obtain, as we should expect to do, the aforesaid equation

$$\frac{x^2}{V^2+\mu c^2} + \frac{y^2}{V^2} = \frac{1}{\mu}.$$

If x, y be the point of contact of (A) with this envelope, *i.e.* the *kinetic focus conjugate* to the point of projection c, 0 in (A), we get

$$\frac{y}{x} = -\frac{b'}{c}\cot a \cdot \frac{V^2}{V^2+\mu c^2}.$$

If a be 0 or 90°, *i.e.* if the point of projection be at the extremity of one of the principal diameters of the ellipse described, we get either $x = 0$ or $y = 0$ at the kinetic focus, shewing that this focus is situated at the extremity of the other principal diameter, which therefore, as will be shewn later, is the last point to which the Action is a minimum.

It is worth remarking that both in this problem and that of the projectile, the direction of motion at the point of contact with the envelope is at right angles to the direction of motion at the point of projection.

We may now draw a very similar series of inferences with reference to this problem to those drawn in the case of the projectile, namely:

If a particle be projected from the given position c, 0 with given velocity V, and it be required to pass through another assigned position, then if this assigned position lie *within* the ellipse

$$\frac{x^2}{V^2+\mu c^2} + \frac{y^2}{V^2} = \frac{1}{\mu}, \quad \ldots \ldots \ldots \ldots \text{(C)}$$

there are four distinct directions in which the particle may be projected so as to pass through the second position, that is to say, two for direct and two for retrograde motion.

If the second position lie *without* the ellipse (C) it will be impossible to project the particle so as to pass through this second position.

If the second position be upon the ellipse (C), there are for the direct motion two coincident directions of projection, and similarly two for the retrograde motion, and the ellipse described by the particle touches (C) at the second position. The ellipse described by the particle always touches (C) either before or after passing through the second position, and the type of the motion, that is to say, the functions of the initial and final coordinates giving the requisite initial and final momenta, and the Action, changes at the point of contact of each trajectory with the envelope (C).

It may be proved, as in case of the projectile, that of two courses to any point P that one which reaches P before touching the envelope has the less Action. But as this is proved subsequently by a general method applicable to all cases, it is unnecessary to verify it in the special case of the ellipse.

There will be another point of contact with the ellipse C, and therefore another kinetic focus and change of type in the second half of the orbit.

Again, when the particle arrives at the extremity of the diameter through the point of projection, $\sqrt{\mu}t = \pi$; that is, $\theta = 2\pi - \theta = \pi$; and the Action there, as is easily seen and will be proved in the sequel, changes type by the adoption of $2\pi - \theta$ instead of θ as the value of $\sqrt{\mu}t$ in its expression. And on again passing through O it changes type by the adoption of $2\pi + \theta$ for $2\pi - \theta$, so that there are in fact four changes of type in each complete revolution, namely, two at the points of contact with the enveloping ellipse, and two at the extremities of the diameter through O.

48.] Analogous propositions to those which we have thus established for special cases can be proved for the general case of any conservative system, having any number of degrees of freedom, acted on by forces continuous functions of the coordinates, and moving from a given initial configuration with the sum of its potential and kinetic energies equal to E.

For let us consider any conservative system with any number, n, of generalised coordinates $q_1, q_2, \ldots q_n$, indicated generally by q, and let this system be acted on by any given forces.

Suppose the system to be initially in any given configuration in which the coordinates are indicated generally by q_0, and to be started from that configuration with total energy E.

Let the initial configuration q_0 be represented by the point O, and the final configuration q by the point P, and let the series of intermediate configurations through which the system passes be represented by the points in the curve OCP; then OCP represents a course or motion of the system from q_0 to q*.

If we attempt, as in the case of the single particle hitherto treated, to express each initial momentum at O in terms of the initial and final coordinates and energy, q_0, q and E, we shall generally find, as we found in case of the particle, that each of these initial momenta will be expressed by a function of the above-mentioned variables having a plurality of forms or values, such as

$$\psi_1(q_0, q, \mathrm{E}), \quad \psi_2(q_0, q, \mathrm{E}) \text{ \&c.},$$

corresponding generally to as many distinct courses or routes by which the system can move from O to P. The time from O to P, as also the Action, will be expressed by functions having a similar plurality of form.

* It will be understood of course that the curve OCP does not represent the motion of the system from the initial to the final configuration in the same way as in the case of the single particle, because each configuration involves many coordinates which cannot be thus graphically denoted. The length of the course must be measured by the time from one configuration to another, as before explained, and is only inadequately represented by the curve joining the points indicating such configurations.

It may be, as in the case of the ellipse before treated, that these functions, or a class of them, although comprehended under one general form, yet contain in their expression a function having many values, as for instance $\cos^{-1} m$, where m is a single-valued function of the coordinates, and differ from each other only by attributing different numerical values to that function. We treat these functions as having different types according to the different values given to the function in question. See Art. 56, *post.*

In the case of the Action, with which we are now chiefly concerned, these forms will be henceforth denoted by

$$f_1(q_0, q, \mathrm{E}), \quad f_2(q_0, q, \mathrm{E}), \text{ \&c.}$$

or shortly f_1, f_2 &c.; and as the initial coordinates q_0, and also E, are supposed invariable, these symbols may be regarded for our present purpose as functions of the final coordinates q only.

49.] It may be that for certain values of the final coordinates, that is for a certain final configuration S, two functions expressing the initial momenta, such as ψ_1, ψ_2, become equal in value for each one of the momenta. In that case two courses from O to S become coincident. The configuration S is then defined to be *a kinetic focus conjugate* to the configuration O.

Inasmuch as there are n initial momenta, this equality gives at first sight n equations for determining the n coordinates of S in order to satisfy the condition. But it must be remembered that, E being given, any one of the initial momenta may be determined as a function of the remaining $n-1$ and E, so that in fact, of the n equations expressing the equality of the initial momenta of the given types ψ_1 and ψ_2, only $n-1$ are independent. They are not then generally sufficient to determine a single position of S, but determine a series of such positions constituting a *quasi locus* or envelope in many respects analogous to the envelope in the cases of the projectile and ellipse. And among other things, this *quasi locus* or envelope has the property that configurations properly situated with regard to it cannot be reached by the system starting from O with momenta of the ψ_1 or ψ_2 type.

Whenever two types, as ψ_1 and ψ_2, become equal in value for every one of the initial momenta, the corresponding courses from O to S become, as above mentioned, coincident, and therefore of course the two corresponding functions expressing the Action become equal in value. But the converse is not true; for two types of the Action, as f_1 and f_2, may for certain final coordinates be equal in value, while the corresponding functions expressing the momenta remain unequal. In that case two non-coincident courses have equal Action from O to the final configuration P.

50.] It appears then that the most general case presents the following analogies with the case of a single particle, viz.

(1) If the final configuration P be arbitrarily chosen, there are generally a certain number (say r) of courses by which the system may move from O to P, these courses being determined by the types of the functions of the coordinates of P selected for expressing the momenta at O.

(2) For certain final configurations any two of these courses may become coincident; for others they may become impossible.

(3) It was proved in the case of the projectile that the function of the final coordinates expressing the Action from the point of projection changes type at the kinetic focus, or point of contact with the envelope. And in like manner, as we proceed to shew, the function expressing the Action from the initial configuration in any conservative system changes type as the system passes through a kinetic focus conjugate to the initial configuration.

(4) It appeared in the case of the ellipse that the Action changes type at the completion of the half period. In like manner we shall shew that if any conservative system, being set in motion, returns by a natural course to the configuration whence it started, making a complete circuit, the Action changes type at the completion of the half circuit.

(5) It was further proved in the case of the projectile that the Action in the natural course from the point of projection O to any point P reached before the change of type is necessarily less than the Action in any infinitely near constrained course from O to P, and is therefore a true minimum, but if P be a point in

the course reached after the change of type, then the Action is not necessarily less in the natural than in the constrained course. Analogous propositions will be proved true for any conservative system.

51.] Let the system move from O in a course $OC_1S\ldots$ (Fig. 8).

Let P be any configuration through which the system passes at the time t. Let the Action in that course from O to P be represented by the function, f_1, of the coordinates of P, so long as P lies between O and a certain configuration S in that course: and at S let $f_1 = f_2$*, f_2 being another of the functions expressing the Action from O. Then if there exist real courses from O to P having Action of the type f_2 for all positions of P between O and S, or C_1 and S, it can be shewn that $f_2 > f_1$ for all such positions of P.

For let OC_2P denote a distinct course from O to P in which the Action has the type f_2.

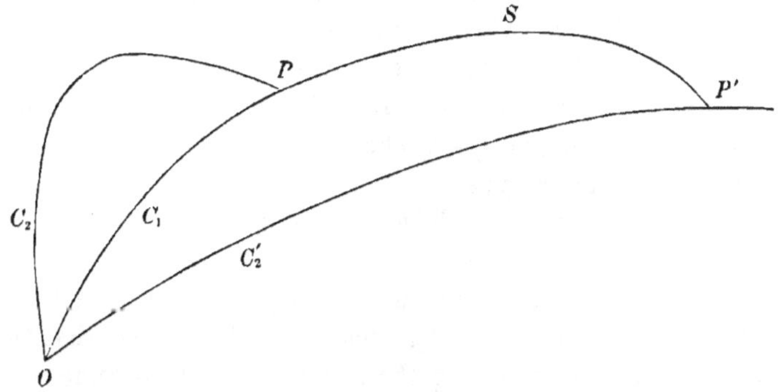

Fig. 8.

Then f_1 and f_2 are both functions of the coordinates of the final configuration P, and as such change with the time t as the system moves on in its course $OC_1P\ldots$. Therefore if $\dot{q}$, p denote the velocities and momenta at P in the course $OC_1P\ldots$, and $\dot{q}'$, p' those in the course OC_2P, we have

$$\frac{d}{dt}(f_1 - f_2) = \Sigma\dot{q}\frac{d}{dq}(f_1 - f_2) = \Sigma\dot{q}p - \Sigma\dot{q}p'.$$

* We use the expression f_1 *at* S as an abbreviation for f_1 *when the final coordinates are those of* S.

Now since E, the total energy, is the same for the two courses OC_1P and OC_2P, the kinetic energy at P is the same in both courses; that is, $\Sigma\dot{q}p = \Sigma\dot{q}'p'$, and therefore $\Sigma\dot{q}p - \Sigma\dot{q}p'$, or $\frac{d}{dt}(f_1 - f_2)$, is necessarily positive (Art. 9, Prop. VIII). Therefore since $f_2 = f_1$ at S, $f_2 > f_1$ if P be reached before S. It thus appears that f_1, the Action *in* the course, always increases faster than f_2, as the system moves on in the course OC_1.... This is a particular case of the theorem of Art. 10.

52.] Next, let P' (Fig. 8) be a configuration in the course OC_1S... beyond S. In that case, remembering the result obtained for the projectile, we do not know whether the Action in OC_1SP' has the type f_1 or f_2, inasmuch as it may change type at S. But whichever type it has, let $OC_2'P'$ denote a course from O to P' in which the Action has the other of the two types in question. Then the process of the last Article shews that Action $OC_1SP' >$ Action $OC_2'P'$, so that Action OC_1SP' has the type f_1 or f_2, whichever is the greater.

We see then that when the system moving in its course OC_1... passes through S, where $f_1 = f_2$, one of two things must happen, viz. either (1) $f_1 - f_2$ changes sign, or (2) the Action in the course changes type.

53.] Three distinct cases have now to be examined.

Firstly, S may be the first kinetic focus conjugate to O in the course OC_1..., and therefore such a configuration that not only two types of Action, f_1 and f_2, become equal, but also that two types of functions expressing the initial momenta become equal when the final coordinates are those of S. In this case there are two coincident courses from O to S.

Or secondly, S may be a configuration at which only two types of the Action become equal, and therefore may be represented by the point of intersection of two non-coincident courses having equal Action (Fig. 9).

Or thirdly, the momenta at S in OC_2S may be respectively equal and opposite to those in OC_1S. In that case the system, in whichever of the two courses it be started from O, returns again to O, so that the two courses are coincident but are described in opposite directions. This case evidently includes all periodic motions.

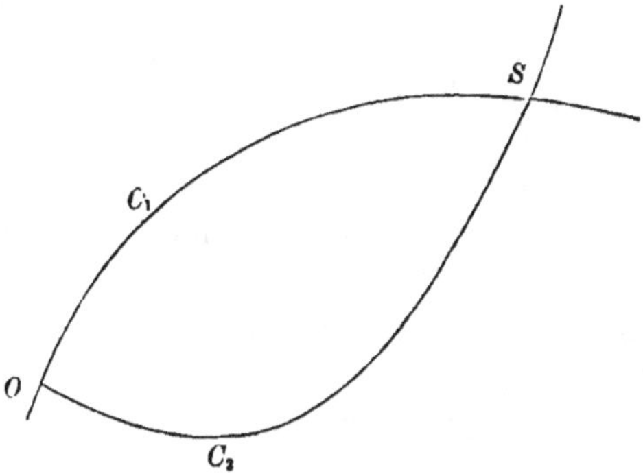

Fig. 9

54.] *To prove that if S be a kinetic focus, the Action must change type at S.*

Let S be a kinetic focus, Q any configuration in OC_1S infinitely near S and beyond it. Then the Action in OC_1SQ has one of the two types f_1, f_2, which become equal at S. Let OC_2QS' be a course in which the Action has the other of those two types.

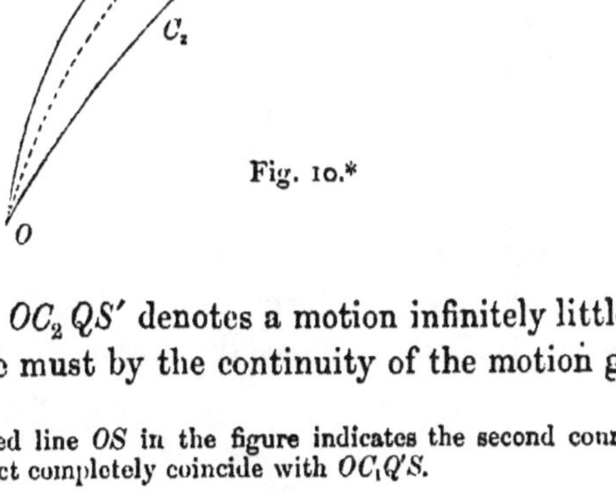

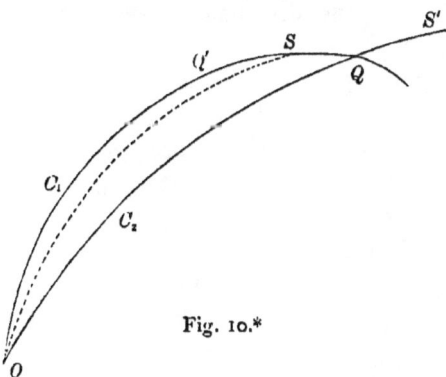

Fig. 10.*

Then since OC_2QS' denotes a motion infinitely little varied from OC_1S, there must by the continuity of the motion generally be a

* The dotted line OS in the figure indicates the second course from O to S. It would in fact completely coincide with $OC_1Q'S$.

kinetic focus corresponding to S somewhere in OC_2QS'. But there can be no such between O and Q: for if there were we could prove by the proposition of Art. 52 that Action OC_2Q is greater than Action OC_1SQ, whereas we have above proved it to be less. The kinetic focus in OC_2QS' must lie at S' beyond Q.

Let Q' be a configuration in OC_1S very near S and between O and S. Then the Action from O to Q in OC_2Q must by the continuity of the motion have the same type as the Action from O to Q' in OC_1Q'. But Action OC_2Q has different type from OC_1Q, for if it had the same the two courses would coincide, which is not the case. Therefore Action OC_1Q' has different type from OC_1Q. And as this is true however near Q and Q' may be to S, provided they do not absolutely coincide with it, and are on opposite sides of it, it follows that the Action in OC_1S must change type at S.

55.] If, on the other hand, S be not a kinetic focus, the two courses OC_1S and OC_2S (Fig. 9) are not coincident, and the momenta at S in OC_1S are not all equal to those in OC_2S. Therefore by Art. 9, Proposition VIII, not only is $\Sigma\dot{q}p-\Sigma\dot{q}p'$ positive, but also it is not zero. That is, $\frac{d}{dt}(f_1-f_2)$ is not zero, and therefore f_1-f_2 generally changes sign at S, and the Action does not change type.

An important exception occurs in the third case above referred to, when the momenta at S in OC_2S are equal and opposite to those in OC_1S, and consequently the system, by whichever of the two courses it be started, returns to O completing the circuit. This case includes all periodic motions. It may be considered as a case of two coincident courses described in opposite directions. We shall find that the Action in such cases changes type at S. For if f_1 be the type which it has at starting, f_1 is zero at the beginning of the circuit, and, being a function of the final coordinates, must also be zero at the end as the system returns to O. Therefore f_1 cannot increase with the time throughout the circuit. But so long as f_1 continues to be the Action in the course, $\frac{df_1}{dt}=2T$, and therefore f_1 must go on increasing with the time at a finite rate. We see then that the Action must change type somewhere in the circuit, and that can only be when $f_1 = f_2$, that is, at S. It must therefore change type at S, and $\frac{df_1}{dt}$ must at that instant change sign discontinuously.

An example of this occurs in the case of the elliptic orbit above discussed, where the Action to S, the extremity of the diameter through the point of projection, has the same value for the retrograde as for the direct motion. And as we said, it there changes type, adopting for the second half of the orbit $2\pi-\theta$ instead of θ in its expression, that is the greater instead of the less value of $\sqrt{\mu t}$ derived from III of Art. 47.

So generally, if the Action in different courses from O to P depends upon the different values of $2i\pi \pm \theta$, where θ is a single-valued function of the coordinates of P, and i an integer, the Action changes type whenever θ is zero or π, although the configuration where that occurs may not be a kinetic focus.

We have thus established the propositions contained in (3) and (4) of Art. 50. It remains to establish that contained in (5).

56.] *If a configuration P be taken on any course OC_1S starting from O, between the configuration O and S, the configuration at which the Action first changes type, the Action from O to P in the course OC_1P will be less than that in any infinitely near constrained course; but if the configuration P be taken beyond S, the Action in the natural course OC_1SP will not necessarily be less than in the infinitely near constrained course.*

For if M be any configuration not in the course but infinitely near some configuration in the course between O and P, it will be always possible by the continuity of the motion for the system to move from O to M by a natural course of the original type, that is, in which the Action is f_1.

At M let the system receive small impulses altering the direction of motion but not the kinetic energy, so as to make it pass to M', and so on by the constrained course $MM'P$ from M to P, being always infinitely near OC_1P.

Then also it will be always possible for the system to move by a natural course OM' having Action of the type f_1, from O to M', any configuration between M and P in the course MP.

Let M' be infinitely near M.

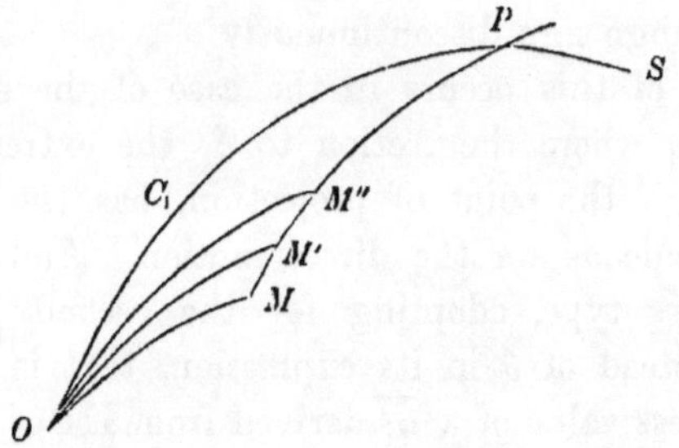

Fig. 11.

Let the coordinates of M be $q_1 \dots q_n$, those of M' $q_1 + \delta q_1$, &c.

Let the action in OM be f, that in OM' $f + \delta f$.

Let p, $\dot{q}$ denote the momenta and velocities in OM, p', $\dot{q}'$ those in MP.

Then,
$$\delta f = \Sigma \frac{df}{dq} \delta q,$$
$$= \Sigma p \delta q,$$
$$= \Sigma p \dot{q}' \delta t,$$

if δt be the time from M to M' in $MM'P$.

But Action $MM' = \Sigma p' \dot{q}' \delta t$. And since the total energy, and therefore also the kinetic energy, is the same at M, in both courses, we have as before
$$\Sigma p' \dot{q}' > \Sigma p \dot{q}';$$

that is, Action OM' — Action $OM <$ Action MM',

or, Action OM + Action $MM' >$ Action OM'.

Similarly, if M'' be any other configuration in $M'P$ infinitely near M',
$$\text{Action } OM' + \text{Action } M'M'' > \text{Action } OM'';$$

and by the continued application of this method we prove that
$$\text{Action } OM + \text{Action } MP > \text{Action } OP;$$

and since M may be as near O as we please, the proof applies to any possible constrained course from O to P, infinitely near OC_1P, and having the total energy E.

The above process would [generally] fail if P were on the other side of S, and therefore the Action in OSP were of a different type from f_1.

For in that case it would, as in the case of the projectile, generally result in proving that Action $(OM+MP)$ is greater, not than Action OC_1P, but than Action $OC_1'P$,* a course from O to P having the same type of Action as OM. And this holds if the change of type consists in giving different values to a multiple function such as $\cos^{-1}m$ (see Art. 48). We should in that case prove that

$$\text{Action } (OM+MP) > \text{Action } OP$$

only where we use both in OM and OP the value of $\cos^{-1}m$ which is less than π, or in both the value between π and 2π, and so on.

Suppose a material system can move from O to P, from P to Q, and from Q to O, the total energy being the same throughout but the momenta at O, P, and Q different in two adjacent courses. If we define OPQ so constituted to be a *kinetic triangle*, the Actions in OP, PQ, and QO its sides; the process of this Article shews that two sides of such a triangle are necessarily greater than the third side, provided the Action in either of the two sides be of the same type with that in the third side; otherwise not necessarily.

57.] We may therefore draw the following conclusions applicable to any conservative system whatever:—

(1) ~~*So long as*~~ *the Action from O* [to] ~~*in*~~ *any* [configuration reached] ~~*course retains the same type,*~~ [before the first kinetic focus] ~~*it*~~ *is less than the Action in any infinitely little varied course from O to the same final configuration, and is therefore a true minimum;*

(2) *But generally ceases to be the least possible when a configuration is passed to which any other course exists having equal Action.*

(3) *After the* ~~*change of type*~~ [first kinetic focus] *the Action ceases to be a minimum.*

* $OC'P$ is not shewn in the figure.

58.] The following is the analogous theorem in Geometry:—

If O be a point on a surface, and geodetic lines be drawn from it; and if S be the point of ultimate intersection of two such geodetics, OC_1S, OC_2S, when they very nearly coincide, then if P be any point in OC_1S between O and S, OC_1P is shorter than any line from O to P that can be drawn upon the surface infinitely near it, but not necessarily shorter than any line whatever drawn on the surface from O to P. If P lie on OC_1S beyond S, a line can be drawn upon the surface from O to P infinitely near to and shorter than OC_1SP.

59.] Now let the system having passed S_1, the first kinetic focus, and there as we have seen acquired f_2 for its type of Action, arrive at a second kinetic focus S_2, where $f_2 = f_3$.

It can then be shewn, exactly as in Art. 54, that the Action again changes type, and becomes f_3 for configurations beyond S_2. And in like manner the system may successively assume all the types $f_1 \ldots f_r$ from the least to the greatest.

60.] As an example of a kinetic focus occurring in a system of many degrees of freedom we may take the case of a system of projectiles. Let, for instance, λ material particles of masses respectively $m_1 \ldots m_\lambda$ be simultaneously projected in the same vertical plane from a point O, the sum of their kinetic energies at O being equal to a given constant E.

Let it be required to find the initial velocities of the several particles in order that the system, so started from O, may pass through a given configuration $x_1, y_1, \ldots x_\lambda, y_\lambda$, that is, that the particle m_1 may be at x_1, y_1, and m_2 at x_2, y_2, and so on all at the same instant. Let O be the origin of coordinates, the axis of x being horizontal in the plane of projection.

Let $u_1 \ldots u_\lambda$ be the horizontal, $v_1 \ldots v_\lambda$ the vertical velocities of the several particles at O. Let t be the time measured from the instant of projection. Then

$$2\,\mathrm{E} = \Sigma m (u^2 + v^2).$$

Also $$u_1 = \frac{x_1}{t}, \quad u_2 = \frac{x_2}{t}, \text{ \&c.};$$

and therefore $$\Sigma m u^2 = \Sigma m \frac{x^2}{t^2}.$$

Also $y_1 = v_1 t - \frac{1}{2} g t^2$, or $v_1 = \frac{y_1}{t} + \frac{1}{2} g t$;

$$\text{similarly } v_2 = \frac{y_2}{t} + \frac{1}{2} g t,$$

$$\&c. = \&c.$$

Therefore $$\Sigma m v^2 = \Sigma m \frac{y^2}{t^2} + \Sigma m \frac{g^2 t^2}{4} + g \Sigma m y.$$

And $$2\,E = \Sigma m \frac{x^2 + y^2}{t^2} + \Sigma m \frac{g^2 t^2}{4} + g \Sigma m y.$$

Whence we obtain

$$\frac{\Sigma m g^2 t^2}{2} = 2\,E - g \Sigma m y \pm \sqrt{4\,E^2 - 4\,E g \Sigma m y - \Sigma m g^2 \Sigma m (x^2 + y^2) + g^2 (\Sigma m y)^2}.$$

It appears then that there are two distinct values, two equal values, or no possible value, of t^2, according as the quantity under the radical sign in the above expression for $\frac{\Sigma m g^2 t^2}{2}$ is positive, zero, or negative.

Again, for any particle, as m_1,

$$\frac{g^2 t^2}{2} = u_1^2 + v_1^2 - g y_1 \pm \sqrt{(u_1^2 + v_1^2)^2 - 2 (u_1^2 + v_1^2) g y_1 - g^2 x_1^2},$$

which for either value of t^2 gives a single value of $u_1^2 + v_1^2$. Similarly

$$\frac{v_1}{u_1} = \frac{y_1 + \frac{1}{2} g t^2}{x_1},$$

which also gives a single value of $\frac{v_1}{u_1}$ for each value of t^2. Therefore for each value of t^2 there is a single possible course for each particle.

It follows that corresponding to the two values of t^2 there are, if the quantity under the radical sign be positive, two, and only two, distinct courses by which the system can pass from O to the given configuration $x_1, y_1 \ldots x_\lambda, y_\lambda$.

When the quantity under the radical sign becomes zero, the two courses become coincident, and the configuration $x_1, y_1 \ldots$ is then a kinetic focus conjugate to O.

Again, by substituting ut for x and $vt - \frac{1}{2} g t^2$ for y in the quantity under the radical, it will be reduced to $(2\,E - g t \Sigma m v)^2$.

Therefore the time at which the kinetic focus is reached is found from $2E - gt\Sigma mv = 0$ or $t = \frac{2E}{g\Sigma mv}$. This is necessarily positive if Σmv be positive, that is if the direction of motion of the centre of gravity of the system at O be above the horizontal line. In that case the system necessarily passes through one and only one kinetic focus after projection from O. The theorems of Arts. 51 and 54 may be verified in this case as in that of the single projectile.

61.] It is evident that a kinetic focus may be regarded as the ultimate intersection of two neighbouring courses from the same initial configuration O, both having the same value of E, the sum of the potential and kinetic energies, when the initial momenta in the one motion differ infinitely little from those in the other.

If the system while in the initial configuration O receive impulses changing its momenta from $p_1 \dots p_n$ to $p_1 + \partial p_1 \dots p_n + \partial p_n$, such impulses are said to constitute a disturbance of the motion; and if the variations of the momenta are such as not to alter the kinetic energy of the system in the given configuration, the disturbance is called a conservative disturbance.

We may obtain a general equation showing the condition that a kinetic focus may exist in a given course from O, and at the same time determining its position if it exist, and the nature of the disturbance in order that the disturbed course may intersect the undisturbed one, i.e. have a configuration in common with it. Thus—

Let $p_1 \dots p_n$ be the initial momenta at O. Then E being given, any one of the p's, e.g. p_n, may be expressed in terms of the others and E, so that only $p_1 \dots p_{n-1}$ are independent.

Let f be the Action in the original system from O to a configuration P whose coordinates are $q_1 \dots q_n$; and let the Action in the varied motion from O to P', whose coordinates are $q_1 + \partial q_1$, &c., have the same value f. Then, p denoting the momenta in the original course at P,

$$\Sigma p \partial q = \Sigma \frac{df}{dq} \partial q = 0.$$

This constitutes one relation between the variations $\partial q_1 \dots \partial q_n$, from which any one, e.g. ∂q_n, may be found in terms of the others, so that only $\delta q_1 \dots \delta q_{n-1}$ are independent.

Now $q_1 \dots q_n$ are functions of $p_1 \dots p_n$ and f, that is, as we have seen, of $p_1 \dots p_{n-1}$, E, and f. If therefore, f being constant, $\frac{dq_1}{dp_1}$ stand for

$$\left(\frac{dq_1}{dp_1}\right) + \frac{dq_1}{dp_n}\frac{dp_n}{dp_1}$$

$\Big(\frac{dp_n}{dp_1}$ being the partial differential coefficient of p_n with regard to p_1 when p_n is expressed as a function of $p_1 \dots p_{n-1}$ and E$\Big)$ and the other coefficients $\frac{dq}{dp}$ in like manner, we have, in order to determine $\partial q_1 \dots \partial q_{n-1}$, the system

$$\left.\begin{aligned} \partial q_1 &= \frac{dq_1}{dp_1}\partial p_1 + \dots + \frac{dq_1}{dp_{n-1}}\partial p_{n-1}, \\ \partial q_2 &= \frac{dq_2}{dp_1}\partial p_1 + \dots + \frac{dq_2}{dp_{n-1}}\partial p_{n-1}, \\ \&c. &= \&c.; \end{aligned}\right\} \dots\dots\dots\dots \text{(A)}$$

in which the coefficients $\frac{dq}{dp}$ are functions of $p_1 \dots p_{n-1}$, E, and f.

In order that P may be a kinetic focus conjugate to O, every ∂q must vanish otherwise than by the vanishing of $\partial p_1 \dots \partial p_{n-1}$, if the proper value of f be substituted in the coefficients $\frac{dq}{dp}$, and proper values given to the ratios of the ∂p's. But this cannot happen for any set of ratios unless the determinant of the system be zero, that is,

$$\Sigma \pm \left(\frac{dq_1}{dp_1} \dots \frac{dq_1}{dp_{n-1}}\right) = 0. \dots\dots\dots\dots\dots\dots \text{(B)}$$

This then is an equation from which f, the action from O to P, and thence also the time t of reaching the kinetic focus, may be determined. If it have one or more real and positive roots differing from zero, each of them corresponds to the time at which the system, started from O with the momenta $p_1 \dots p_{n-1}$, reaches a kinetic focus conjugate to O, and therefore determines the position of that focus.

In the case of the projectile or any other motion in a plane curve, the curve of equal Action is normal to the course. If the undisturbed course intersect it in P, and the disturbed one in P', then let $PP' = \partial q$, or if ∂p be the variation of one of the two initial momenta, $PP' = \frac{dq}{dp}\partial p$.

In order that P may be a kinetic focus we must have $\frac{dq}{dp} = 0$, that being the form which the equation (B) assumes in this case.

Further, in case there be more than two degrees of freedom, if the equation (B) be satisfied, it harmonises the equations (A), and they then suffice to determine the ratios which $\partial p_1 \,..\, \partial p_{n-1}$ must bear to each other, that is the particular kind of disturbance, in order that the disturbed and undisturbed courses may have a configuration in common.

If for different roots of the equation (B) these ratios have different values, they correspond to distinct disturbed courses from O, each of which intersects the original course. If two or more roots of (B) correspond to one and the same set of ratios of $\partial p_1 \ldots \partial p_{n-1}$, then the same disturbed course intersects the undisturbed course more than once. The second and subsequent intersections may with propriety be called secondary kinetic foci. In the elliptic orbit before considered, the disturbed course intersects the undisturbed one four times in each complete revolution. In the case of the projectile, on the other hand, the two courses, having once intersected each other in the kinetic focus, will never after again intersect.

If in equation (B) we were to substitute for $p_1 \ldots p_{n-1}$ and f their values as functions of the initial and final coordinates, (B) would then be an equation between the final coordinates involving the initial coordinates as constants; the equation namely to the locus of kinetic foci, or envelope of the system.

CHAPTER V.

APPENDIX.

ARTICLE 62.] We referred in Article 1 to an expression for p_r, the generalised component of momentum corresponding to the coordinate q_r in the language of quaternions, viz. the scalar function

$$-\Sigma' m\, S\frac{d\rho}{dt}\frac{d\rho}{dq_r},$$

ρ being the vector from the origin to an element of the system of mass m, expressed as a function of the n scalar variables $q_1 \dots q_n$, and Σ' denoting summation for all such elements. In like manner

$$-\Sigma' m\, S\frac{d^2\rho}{dt^2}\frac{d\rho}{dq_r}$$

denotes the generalised component of force, F_r, corresponding to the coordinate q_r.

If we denote by P_r and G_r respectively the corresponding vector functions

$$\Sigma' m V\frac{d\rho}{dt}\frac{d\rho}{dq_r} \quad \text{and} \quad \Sigma' m V\frac{d^2\rho}{dt^2}\frac{d\rho}{dq_r},$$

it will be found that P and G possess analytical properties similar in many respects to those already investigated for p and F. For we have

$$\frac{d\rho}{dt} = \frac{d\rho}{dq_1}\dot{q}_1 + \frac{d\rho}{dq_2}\dot{q}_2 + \dots + \frac{d\rho}{dq_n}\dot{q}_n;$$

substituting which in the expressions for p and P respectively, we obtain

$$-p_r = \Sigma' m S \frac{d\rho}{dq_1}\frac{d\rho}{dq_r}\dot{q}_1 + \ldots + \Sigma' m S (\frac{d\rho}{dq_r})^2 \dot{q}_r + \ldots + \Sigma' m S \frac{d\rho}{dq_n}\frac{d\rho}{dq_r}\dot{q}_n,$$

$$P_r = \Sigma' m V \frac{d\rho}{dq_1}\frac{d\rho}{dq_r}\dot{q}_1 + \text{- - - - - - - - - - - - -} + \Sigma' m V \frac{d\rho}{dq_n}\frac{d\rho}{dq_r}\dot{q}_n;$$

in which every $\frac{d\rho}{dq}$ is a vector. Evidently in the expression for P_r the term involving $\dot{q}_r$ disappears, because

$$V(\frac{d\rho}{dq_r})^2 = 0.$$

If we write these equations in the form

$$-p_r = a_{1r}\dot{q}_1 + \ldots + a_{rr}\dot{q}_r + \ldots + a_{nr}\dot{q}_n,$$
$$P_r = b_{1r}\dot{q}_1 + \ldots \text{- - - - - - -} + b_{nr}\dot{q}_n,$$

the coefficients a are all scalars, and the coefficients b are the corresponding vectors.

Further,

$$a_{rs} = \Sigma' m S \frac{d\rho}{dq_r}\frac{d\rho}{dq_s} = \Sigma' m S \frac{d\rho}{dq_s}\frac{d\rho}{dq_r} = a_{sr},$$

as already proved, but

$$b_{rs} = \Sigma' m V \frac{d\rho}{dq_r}\frac{d\rho}{dq_s} = -\Sigma' m V \frac{d\rho}{dq_s}\frac{d\rho}{dq_r} = -b_{sr}.$$

If $\dot{q}'_1 \ldots \dot{q}'_n$ be any other set of component velocities which the system might have in the same configuration, we shall obtain from the above

$$\Sigma p\dot{q}' = \Sigma p'\dot{q},$$

as above proved, but

$$\Sigma P\dot{q}' = -\Sigma P'\dot{q},$$

Σ denoting summation from 1 to n. Also

$$\Sigma P\dot{q}' = b_{12}\{\dot{q}_2\dot{q}'_1 - q_1\dot{q}'_2\} + b_{13}\{\dot{q}_3\dot{q}'_1 - \dot{q}_1\dot{q}'_3\} + \ldots$$

and by making $\dot{q}' = \dot{q}$, we obtain

$$\Sigma P\dot{q} = 0.$$

Again, Lagrange's equations may be written

$$\frac{dp}{dt} = \tfrac{1}{2}\Sigma\dot{q}\frac{dp}{dq} + F_q = \tfrac{1}{2}\Sigma\dot{q}\frac{dp}{dq} - \Sigma' m S \frac{d^2\rho}{dt^2}\frac{d\rho}{dq},$$

to which corresponds

$$\frac{dP}{dt} = \tfrac{1}{2}\Sigma\dot{q}\frac{dP}{dq} + \Sigma' m V \frac{d^2\rho}{dt^2}\frac{d\rho}{dq} = \tfrac{1}{2}\Sigma\dot{q}\frac{dP}{dq} + G_q.$$

63.] The scalar function $-\Sigma' m S\frac{d\rho}{dt}\rho$ is equal to

$$\Sigma' m\left(x\frac{dx}{dt}+y\frac{dy}{dt}+z\frac{dz}{dt}\right) \text{ or } \tfrac{1}{2}\frac{d}{dt}\Sigma' m(x^2+y^2+z^2).$$

$\Sigma' m V\frac{d\rho}{dt}\rho$ is twice the vector area described by the system per unit of time about an axis through the origin.

In Cartesian coordinates $\Sigma' m\left(x\frac{dy}{dt}-y\frac{dx}{dt}\right)$ is twice the area described per unit of time about the axis of z. If a line be measured off along that axis representing in magnitude

$$\Sigma' m\left\{x\frac{dy}{dt}-y\frac{dx}{dt}\right\},$$

and if the corresponding lines be taken for the other two axes, and the resultant of these three lines be formed, that resultant is represented in magnitude and direction by $\Sigma' m V\frac{d\rho}{dt}\rho$.

If this be denoted by A, and if the corresponding scalar $-\Sigma' m S\frac{d\rho}{dt}\rho$, or its equivalent $\Sigma' m\left\{x\frac{dx}{dt}+y\frac{dy}{dt}+z\frac{dz}{dt}\right\}$, be denoted by S, we see that P stands in the same relation to A in which p stands to S, so that

$$p=\frac{dS}{dq} \text{ and } P=\frac{dA}{dq},$$

the actual velocities $\dot{x}$, $\dot{y}$, $\dot{z}$ being in either case regarded as constant.

64.] We conclude with Clausius' theorem on the equality of the mean kinetic energy and the mean virial, as expressed in generalised coordinates. In the expressions obtained in the last Article, if for the linear velocities

$$\frac{d\rho}{dt}, \text{ or } \frac{dx}{dt}, \frac{dy}{dt}, \frac{dz}{dt},$$

we substitute the effective accelerations

$$\frac{d^2\rho}{dt^2}, \text{ or } \frac{d^2x}{dt^2}, \frac{d^2y}{dt^2}, \frac{d^2z}{dt^2},$$

the scalar S becomes

$$-\Sigma' m S\frac{d^2\rho}{dt^2}\rho \text{ or } \Sigma' m\{Xx+Yy+Zz\},$$

the half of which is called the *virial* of the system, which we will denote by V. The vector A becomes the moment of the resultant couple.

Now let us suppose that the nature of our material system is such that the mean value of $\Sigma' m \frac{d\rho}{dt}\rho$ is constant, if the time for which the mean is taken be sufficiently great. That is evidently the case for every strictly periodic motion if the means be taken for the periodic time; and it may be the case for motions which are not strictly periodic if a sufficiently long time be in question. Any such motion may be defined to be stationary.

As the expression $\Sigma' m \frac{d\rho}{dt}\rho$ has both a scalar and a vector part, both must be separately constant on average of the time in question, or, which is the same thing, both

$$\left.\begin{array}{ll} & \frac{d}{dt}\Sigma' m S \frac{d\rho}{dt}\rho = 0 \text{ on average.} \\ \text{and} & \frac{d}{dt}\Sigma' m V \frac{d\rho}{dt}\rho = 0 \text{ on average.} \end{array}\right\} \quad \cdots\cdots\cdots\cdots \text{(C)}$$

The first of these equations gives

$$\Sigma' m S \frac{d^2\rho}{dt^2}\rho + \Sigma' m \left(\frac{d\rho}{dt}\right)^2 = 0,$$

or the mean kinetic energy added to the mean virial is zero.

The second of equations (C) expresses the principle of conservation of areas.

65.] If now ρ be such a function of $q_1 \ldots q_n$ as that

$$\rho = \Sigma q \frac{d\rho}{dq},$$

$$\text{then} \qquad -\Sigma' m S \frac{d\rho}{dt}\rho = \Sigma p q,$$

$$\text{and} \qquad \Sigma' m V \frac{d\rho}{dt}\rho = \Sigma P q,$$

so that in stationary motion both Σpq and ΣPq are constant on average. And in this case

$$\begin{aligned} \Sigma F q &= -\Sigma' m S \frac{d^2\rho}{dt^2} \Sigma q \frac{d\rho}{dq}, \\ &= -\Sigma' m S \frac{d^2\rho}{dt^2}\rho = \Sigma' m (Xx + Yy + Zz), \end{aligned}$$

and is therefore identical with the virial as hitherto defined. We have then in this case $\Sigma Fq + 2T = 0$.

66.] In the general case an analogous theorem to that of the last article may be proved thus :—

If Σpq be constant on average, then on average

$$\frac{d}{dt}\Sigma pq = 0;$$

that is, $$\Sigma \dot{q}\frac{dp}{dt} + \Sigma p\dot{q} = 0;$$

that is, $$\Sigma q\left(\frac{dT}{dq} - \frac{dU}{dq}\right) + 2T = 0;$$

or, writing L for $T - U$, remembering that

$$p = \frac{dT}{d\dot{q}} = \frac{dL}{d\dot{q}},$$

$$\Sigma q\frac{dL}{dq} + \Sigma \dot{q}\frac{dL}{d\dot{q}} = 0.$$

Clarendon Press, Oxford.

BOOKS

PUBLISHED FOR THE UNIVERSITY BY

MACMILLAN AND CO., LONDON;

ALSO TO BE HAD AT THE

CLARENDON PRESS DEPOSITORY, OXFORD.

LEXICONS, GRAMMARS, &c.

(*See also Clarendon Press Series* pp. 21, 24.)

A Greek-English Lexicon, by Henry George Liddell, D.D., and Robert Scott, D.D. *Sixth Edition, Revised and Augmented.* 1870. 4to. *cloth*, 1*l*. 16*s*.

A Greek-English Lexicon, abridged from the above, chiefly for the use of Schools. *Seventeenth Edition. Carefully Revised throughout.* 1876. Square 12mo. *cloth*, 7*s*. 6*d*.

A copious Greek-English Vocabulary, compiled from the best authorities. 1850. 24mo. *bound*, 3*s*.

Graecae Grammaticae Rudimenta in usum Scholarum. Auctore Carolo Wordsworth, D.C.L. *Eighteenth Edition*, 1875. 12mo. *bound*, 4*s*.

A Greek Primer, for the use of beginners in that Language. By the Right Rev. Charles Wordsworth, D.C.L., Bishop of St. Andrews. *Sixth Edition, Revised and Enlarged.* Extra fcap. 8vo. *cloth*, 1*s*. 6*d*.

A Practical Introduction to Greek Accentuation, by H. W. Chandler, M.A. 1862. 8vo. *cloth*, 10*s*. 6*d*.

Etymologicon Magnum. Ad Codd. MSS. recensuit et notis variorum instruxit Thomas Gaisford, S.T.P. 1848. fol. *cloth*, 1*l*. 12*s*.

Suidae Lexicon. Ad Codd. MSS. recensuit Thomas Gaisford, S.T.P. Tomi III. 1834. fol. *cloth*, 2*l*. 2*s*.

Scheller's Lexicon of the Latin Tongue, with the German explanations translated into English by J. E. Riddle, M.A. 1835. fol. *cloth*, 1*l*. 1*s*.

Scriptores Rei Metricae. Edidit Thomas Gaisford, S.T.P. Tomi III. 8vo. *cloth*, 15*s*.

Sold separately:

Hephaestion, Terentianus Maurus, Proclus, cum annotationibus, etc. Tomi II. 10*s*. Scriptores Latini. 5*s*.

The Book of Hebrew Roots, by Abu 'L-Walîd Marwân ibn Janâh, otherwise called Rabbî Yônâh. Now first edited, with an Appendix, by Ad. Neubauer. 4to. *cloth*, 2*l*. 7*s*. 6*d*.

A Treatise on the use of the Tenses in Hebrew. By S. R. Driver, M.A. Extra fcap. 8vo. *cloth*, 6*s*. 6*d*.

Thesaurus Syriacus: collegerunt Quatremère, Bernstein, Lorsbach, Arnoldi, Field: edidit R. Payne Smith, S.T.P.R.

Fasc. I–IV. 1868–77. sm. fol. *each*, 1*l*. 1*s*.

Lexicon Aegyptiaco-Latinum ex veteribus Linguae Aegyptiacae Monumentis, etc., cum Indice Vocum Latinarum ab H. Tattam, A.M. 1835. 8vo. *cloth*, 15*s*.

A Practical Grammar of the Sanskrit Language, arranged with reference to the Classical Languages of Europe, for the use of English Students, by Monier Williams, M.A. *Fourth Edition*, 1877. 8vo. *cloth*, 15*s*.

A Sanskrit-English Dictionary, Etymologically and Philologically arranged, with special reference to Greek, Latin, German, Anglo-Saxon, English, and other cognate Indo-European Languages. By Monier Williams, M.A., Boden Professor of Sanskrit. 1872. 4to. *cloth*, 4*l*. 14*s*. 6*d*.

Nalopákhyánam. Story of Nala, an Episode of the Mahá-Bhárata: the Sanskrit text, with a copious Vocabulary, Grammatical Analysis, and Introduction, by Monier Williams, M.A. The Metrical Translation by the Very Rev. H. H. Milman, D.D. 1860. 8vo. *cloth*, 15*s*.

Sakuntalā. A Sanskrit Drama, in seven Acts. Edited by Monier Williams, M.A., D.C.L., Boden Professor of Sanskrit. *Second Edition*, 8vo. *cloth*, 21*s*.

An Anglo-Saxon Dictionary, by Joseph Bosworth, D.D., Professor of Anglo-Saxon, Oxford. *New edition. In the Press.*

An Icelandic-English Dictionary. Based on the MS. collections of the late Richard Cleasby. Enlarged and completed by G. Vigfússon. With an Introduction, and Life of Richard Cleasby, by G. Webbe Dasent, D.C.L. 4to. *cloth*, 3*l*. 7*s*.

A List of English Words the Etymology of which is illustrated by comparison with Icelandic. Prepared in the form of an Appendix to the above. By W. W. Skeat, M.A., *stitched*, 2*s*.

A Handbook of the Chinese Language. Parts I and II, Grammar and Chrestomathy. By James Summers. 1863. 8vo. *half bound*, 1*l*. 8*s*.

Cornish Drama (The Ancient). Edited and translated by E. Norris, Esq., with a Sketch of Cornish Grammar, an Ancient Cornish Vocabulary, etc. 2 vols. 1859. 8vo. *cloth*, 1*l*. 1*s*.

The Sketch of Cornish Grammar separately, *stitched*, 2*s*. 6*d*.

An Etymological Dictionary of the English Language, arranged on an Historical Basis. By W. W. Skeat, M.A., Elrington and Bosworth Professor of Anglo-Saxon in the University of Cambridge. To be completed in Four Parts. Part I, containing A—D, *just ready*.

GREEK CLASSICS, &c.

Aeschylus: quae supersunt in Codice Laurentiano typis descripta. Edidit R. Merkel. 1861. Small folio, *cloth*, 1*l*. 1*s*.

Aeschylus: Tragoediae et Fragmenta, ex recensione Guil. Dindorfii. *Second Edition*, 1851. 8vo. *cloth*, 5*s*. 6*d*.

Aeschylus: Annotationes Guil. Dindorfii. Partes II. 1841. 8vo. *cloth*, 10*s*.

Aeschylus: Scholia Graeca, ex Codicibus aucta et emendata a Guil. Dindorfio. 1851. 8vo. *cloth*, 5*s*.

Sophocles: Tragoediae et Fragmenta, ex recensione et cum commentariis Guil. Dindorfii. *Third Edition*, 2 vols. 1860. fcap. 8vo. *cloth*, 1*l*. 1*s*.

Each Play separately, *limp*, 2*s*. 6*d*.

The Text alone, printed on writing paper, with large margin, royal 16mo. *cloth*, 8*s*.

The Text alone, square 16mo. *cloth*, 3*s*. 6*d*.

Each Play separately, *limp*, 6*d*. (See also p. 26.)

Sophocles: Tragoediae et Fragmenta cum Annotatt. Guil. Dindorfii. Tomi II. 1849. 8vo. *cloth*, 10*s*.

The Text, Vol. I. 5*s*. 6*d*. The Notes, Vol. II. 4*s*. 6*d*.

Sophocles: Scholia Graeca:

Vol. I. ed. P. Elmsley, A.M. 1825. 8vo. *cloth*, 4*s*. 6*d*.
Vol. II. ed. Guil. Dindorfius. 1852. 8vo. *cloth*, 4*s*. 6*d*.

Euripides: Tragoediae et Fragmenta, ex recensione Guil. Dindorfii. Tomi II. 1834. 8vo. *cloth*, 10*s*.

Euripides: Annotationes Guil. Dindorfii. Partes II. 1840. 8vo. *cloth*, 10*s*.

Euripides: Scholia Graeca, ex Codicibus aucta et emendata a Guil. Dindorfio. Tomi IV. 1863. 8vo. *cloth*, 1*l.* 16*s.*

Euripides: Alcestis, ex recensione Guil. Dindorfii. 1834. 8vo. *sewed*, 2*s.* 6*d.*

Aristophanes: Comoediae et Fragmenta, ex recensione Guil. Dindorfii. Tomi II. 1835. 8vo. *cloth*, 11*s.*

Aristophanes: Annotationes Guil. Dindorfii. Partes II. 1837. 8vo. *cloth*, 11*s.*

Aristophanes: Scholia Graeca, ex Codicibus aucta et emendata a Guil. Dindorfio. Partes III. 1839. 8vo. *cloth*, 1*l.*

Aristophanem, Index in: J. Caravellae. 1822. 8vo. *cloth*, 3*s.*

Metra Aeschyli Sophoclis Euripidis et Aristophanis. Descripta a Guil. Dindorfio. Accedit Chronologia Scenica. 1842. 8vo. *cloth*, 5*s.*

Anecdota Graeca Oxoniensia. Edidit J. A. Cramer, S.T.P. Tomi IV. 8vo. *cloth*, 1*l.* 2*s.*

Anecdota Graeca e Codd. MSS. Bibliothecae Regiae Parisiensis. Edidit J. A. Cramer, S.T.P. Tomi IV. 8vo. *cloth*, 1*l.* 2*s.*

Apsinis et Longini Rhetorica. E Codicibus MSS. recensuit Joh. Bakius. 1849. 8vo. *cloth*, 3*s.*

Aristoteles; ex recensione Immanuelis Bekkeri. Accedunt Indices Sylburgiani. Tomi XI. 1837. 8vo. *cloth*, 2*l.* 10*s.*

The volumes (except vol. IX.) may be had separately, price 5*s.* 6*d.* each.

Aristotelis Ethica Nicomachea, ex recensione Immanuelis Bekkeri. Crown 8vo. *cloth*, 5*s.*

Choerobosci Dictata in Theodosii Canones, necnon Epimerismi in Psalmos. E Codicibus MSS. edidit Thomas Gaisford, S.T.P. Tomi III. 1842. 8vo. *cloth*, 15*s.*

Demosthenes: ex recensione Guil. Dindorfii. Tomi I. II. III. IV. 1846. 8vo. *cloth*, 1*l.* 1*s.*

Demosthenes: Tomi V. VI. VII. Annotationes Interpretum. 1849. 8vo. *cloth*, 15*s.*

Demosthenes: Tomi VIII. IX. Scholia. 1851. 8vo. *cloth*, 10*s.*

Harpocrationis Lexicon, ex recensione G. Dindorfii. Tomi II. 1854. 8vo. *cloth*, 10*s.* 6*d.*

Heracliti Ephesii Reliquiae. Recensuit I. Bywater, M.A. 8vo. *cloth*, price 6*s.*

Herculanensium Voluminum Partes II. 8vo. *cloth*, 10*s.*

Homerus: Ilias, cum brevi Annotatione C. G. Heynii. Accedunt Scholia minora. Tomi II. 1834. 8vo. *cloth*, 15*s.*

Homerus: Ilias, ex rec. Guil. Dindorfii. 1856. 8vo. *cloth*, 5*s.* 6*d.*

Homerus: Scholia Graeca in Iliadem. Edited by Prof. W. Dindorf, after a new collation of the Venetian MSS. by D. B. Monro, M.A., Fellow of Oriel College.
Vols. I. II. 8vo. *cloth*, 24*s*. Vols. III. IV. 8vo. *cloth*, 26*s*.

Homerus: Odyssea, ex rec. Guil. Dindorfii. 8vo. *cloth*, 5*s*. 6*d*.

Homerus: Scholia Graeca in Odysseam. Edidit Guil. Dindorfius. Tomi II. 1855. 8vo. *cloth*, 15*s*. 6*d*.

Homerum, Index in: Seberi. 1780. 8vo. *cloth*, 6*s*. 6*d*.

Oratores Attici ex recensione Bekkeri:

I. Antiphon, Andocides, et Lysias. 1822. 8vo. *cloth*, 7*s*.
II. Isocrates. 1822. 8vo. *cloth*, 7*s*.
III. Isaeus, Aeschines, Lycurgus, Dinarchus, etc. 1823. 8vo. *cloth*, 7*s*.

Scholia Graeca in Aeschinem et Isocratem. Edidit G. Dindorfius. 1852. 8vo. *cloth*, 4*s*.

Paroemiographi Graeci, quorum pars nunc primum ex Codd. MSS. vulgatur. Edidit T. Gaisford, S.T.P. 1836. 8vo. *cloth*, 5*s*. 6*d*.

Plato: The Apology, with a revised Text and English Notes, and a Digest of Platonic Idioms, by James Riddell, M.A. 1867. 8vo. *cloth*, 8*s*. 6*d*.

Plato: Philebus, with a revised Text and English Notes, by Edward Poste, M.A. 1860. 8vo. *cloth*, 7*s*. 6*d*.

Plato: Sophistes and Politicus, with a revised Text and English Notes, by L. Campbell, M.A. 1866. 8vo. *cloth*, 18*s*.

Plato: Theaetetus, with a revised Text and English Notes, by L. Campbell, M.A. 1861. 8vo. *cloth*, 9*s*.

Plato: The Dialogues, translated into English, with Analyses and Introductions, by B. Jowett, M.A., Regius Professor of Greek. *A new Edition in* 5 *volumes*, medium 8vo. *cloth*, 3*l*. 10*s*.

Plato: Index to. Compiled for the Second Edition of Professor Jowett's Translation of the Dialogues. By Evelyn Abbott, M.A., Fellow and Tutor of Balliol College. Demy 8vo. *paper covers*, 2*s*. 6*d*.

Plato: The Republic, with a revised Text and English Notes, by B. Jowett, M.A., Regius Professor of Greek. Demy 8vo. *Preparing*.

Plotinus. Edidit F. Creuzer. Tomi III. 1835. 4to. 1*l*. 8*s*.

Stobaei Florilegium. Ad MSS. fidem emendavit et supplevit T. Gaisford, S.T.P. Tomi IV. 8vo. *cloth*, 1*l*.

Stobaei Eclogarum Physicarum et Ethicarum libri duo. Accedit Hieroclis Commentarius in aurea carmina Pythagoreorum. Ad MSS. Codd. recensuit T. Gaisford, S.T.P. Tomi II. 8vo. *cloth*, 11*s*.

Xenophon: Historia Graeca, ex recensione et cum annotationibus L. Dindorfii. *Second Edition*, 1852. 8vo. *cloth*, 10s. 6d.

Xenophon: Expeditio Cyri, ex rec. et cum annotatt. L. Dindorfii. *Second Edition*, 1855. 8vo. *cloth*, 10s. 6d.

Xenophon: Institutio Cyri, ex rec. et cum annotatt. L. Dindorfii. 1857. 8vo. *cloth*, 10s. 6d.

Xenophon: Memorabilia Socratis, ex rec. et cum annotatt. L. Dindorfii. 1862. 8vo. *cloth*, 7s. 6d.

Xenophon: Opuscula Politica Equestria et Venatica cum Arriani Libello de Venatione, ex rec. et cum annotatt. L. Dindorfii. 1866. 8vo. *cloth*, 10s. 6d.

THE HOLY SCRIPTURES, &c.

The Holy Bible in the earliest English Versions, made from the Latin Vulgate by John Wycliffe and his followers: edited by the Rev. J. Forshall and Sir F. Madden. 4 vols. 1850. royal 4to. *cloth*, 3l. 3s.

The Holy Bible: an exact reprint, page for page, of the Authorized Version published in the year 1611. Demy 4to. *half bound*, 1l. 1s.

Vetus Testamentum Graece cum Variis Lectionibus. Editionem a R. Holmes, S.T.P. inchoatam continuavit J. Parsons, S.T.B. Tomi V. 1798–1827. folio, 7l.

Vetus Testamentum ex Versione Septuaginta Interpretum secundum exemplar Vaticanum Romae editum. Accedit potior varietas Codicis Alexandrini. Tomi III. *Editio Altera.* 18mo. *cloth*, 18s.

Origenis Hexaplorum quae supersunt; sive, Veterum Interpretum Graecorum in totum Vetus Testamentum Fragmenta. Edidit Fridericus Field, A.M. 2 vols. 1867–1874. 4to. *cloth*, 5l. 5s.

Libri Psalmorum Versio antiqua Latina, cum Paraphrasi Anglo-Saxonica. Edidit B. Thorpe, F.A.S. 1835. 8vo. *cloth*, 10s. 6d.

Libri Psalmorum Versio antiqua Gallica e Cod. MS. in Bibl. Bodleiana adservato, una cum Versione Metrica aliisque Monumentis pervetustis. Nunc primum descripsit et edidit Franciscus Michel, Phil. Doct. 1860. 8vo. *cloth*, 10s. 6d.

Libri Prophetarum Majorum, cum Lamentationibus Jeremiae, in Dialecto Linguae Aegyptiacae Memphitica seu Coptica. Edidit cum Versione Latina H. Tattam, S.T.P. Tomi II. 1852. 8vo. *cloth*, 17s.

Libri duodecim Prophetarum Minorum in Ling. Aegypt. vulgo Coptica. Edidit H. Tattam, A.M. 1836. 8vo. *cloth*, 8s. 6d.

Novum Testamentum Graece. Antiquissimorum Codicum Textus in ordine parallelo dispositi. Accedit collatio Codicis Sinaitici. Edidit E. H. Hansell, S.T.B. Tomi III. 1864. 8vo. *half morocco*, 2*l.* 12*s.* 6*d.*

Novum Testamentum Graece. Accedunt parallela S. Scripturae loca, necnon vetus capitulorum notatio et canones Eusebii. Edidit Carolus Lloyd, S.T.P.R., necnon Episcopus Oxoniensis. 1876. 18mo. *cloth*, 3*s.*

The same on writing paper, with large margin, cloth, 10*s.* 6*d.*

Novum Testamentum Graece juxta Exemplar Millianum. 1876. 18mo. *cloth*, 2*s.* 6*d.*

The same on writing paper, with large margin, cloth, 9*s.*

Evangelia Sacra Graecae. fcap. 8vo. *limp*, 1*s.* 6*d.*

The New Testament in Greek and English, on opposite pages, arranged and edited by E. Cardwell, D.D. 2 vols. 1837. crown 8vo. *cloth*, 6*s.*

Novum Testamentum Coptice, cura D. Wilkins. 1716. 4to. *cloth*, 12*s.* 6*d.*

Evangeliorum Versio Gothica, cum Interpr. et Annott. E. Benzelii. Edidit, et Gram. Goth. praemisit, E. Lye, A.M. 1759. 4to. *cloth*, 12*s.* 6*d.*

Diatessaron; sive Historia Jesu Christi ex ipsis Evangelistarum verbis apte dispositis confecta. Ed. J. White. 1856. 12mo. *cloth*, 3*s.* 6*d.*

Canon Muratorianus. The earliest Catalogue of the Books of the New Testament. Edited with Notes and a Facsimile of the MS. in the Ambrosian Library at Milan, by S. P. Tregelles, LL.D. 1868. 4to. *cloth*, 10*s.* 6*d.*

The Five Books of Maccabees, in English, with Notes and Illustrations by Henry Cotton, D.C.L. 1833. 8vo. *cloth*, 10*s.* 6*d.*

Horae Hebraicae et Talmudicae, a J. Lightfoot. *A new Edition*, by R. Gandell, M.A. 4 vols. 1859. 8vo. *cloth*, 1*l.* 1*s.*

FATHERS OF THE CHURCH, &c.

Liturgies, Eastern and Western: being a Reprint of the Texts, either original or translated, of the most representative Liturgies of the Church, from various sources. Edited, with Introduction, Notes, and a Liturgical Glossary, by C. E. Hammond, M.A., author of Textual Criticism applied to the New Testament. Crown 8vo. *cloth*, 10*s.* 6*d.*

Athanasius: The Orations of St. Athanasius against the Arians. With an Account of his Life. By William Bright, D.D., Regius Professor of Ecclesiastical History, Oxford. Crown 8vo. *cloth*, 9*s.*

The Canons of the First Four General Councils of Nicaea, Constantinople, Ephesus, and Chalcedon. Crown 8vo. *cloth*, 2*s.* 6*d.*

Catenae Graecorum Patrum in Novum Testamentum. Edidit J. A. Cramer, S.T.P. Tomi VIII. 1838–1844. 8vo. *cloth*, 2*l*. 4*s*.

Clementis Alexandrini Opera, ex recensione Guil. Dindorfii. Tomi IV. 1869. 8vo. *cloth*, 3*l*.

Cyrilli Archiepiscopi Alexandrini in XII Prophetas. Edidit P. E. Pusey, A.M. Tomi II. 1868. 8vo. *cloth*, 2*l*. 2*s*.

Cyrilli Archiepiscopi Alexandrini in D. Joannis Evangelium. Accedunt Fragmenta Varia necnon Tractatus ad Tiberium Diaconum Duo. Edidit post Aubertum P. E. Pusey, A.M. Tomi III. 8vo. *cloth*, 2*l*. 5*s*.

Cyrilli Archiepiscopi Alexandrini Commentarii in Lucae Evangelium quae supersunt Syriace. E MSS. apud Mus. Britan. edidit R. Payne Smith, A.M. 1858. 4to. *cloth*, 1*l*. 2*s*.

The same, translated by R. Payne Smith, M.A. 2 vols. 1859. 8vo. *cloth*, 14*s*.

Ephraemi Syri, Rabulae Episcopi Edesseni, Balaei, aliorumque. Opera Selecta. E Codd. Syriacis MSS. in Museo Britannico et Bibliotheca Bodleiana asservatis primus edidit J. J. Overbeck. 1865. 8vo. *cloth*, 1*l*. 1*s*.

Eusebii Pamphili Evangelicae Praeparationis Libri XV. Ad Codd. MSS. recensuit T. Gaisford, S.T.P. Tomi IV. 1843. 8vo. *cloth*, 1*l*. 10*s*.

Eusebii Pamphili Evangelicae Demonstrationis Libri X. Recensuit T. Gaisford, S.T.P. Tomi II. 1852. 8vo. *cloth*, 15*s*.

Eusebii Pamphili contra Hieroclem et Marcellum Libri. Recensuit T. Gaisford, S.T.P. 1852. 8vo. *cloth*, 7*s*.

Eusebius' Ecclesiastical History, according to the text of Burton. With an Introduction by William Bright, D.D. Crown 8vo. *cloth*, 8*s*. 6*d*.

Eusebii Pamphili Hist. Eccl.: **Annotationes Variorum.** Tomi II. 1842. 8vo. *cloth*, 17*s*.

Evagrii Historia Ecclesiastica, ex recensione H. Valesii. 1844. 8vo. *cloth*, 4*s*.

Irenaeus: The Third Book of St. Irenaeus, Bishop of Lyons, against Heresies. With short Notes, and a Glossary. By H. Deane, B.D., Fellow of St. John's College, Oxford. Crown 8vo. *cloth*, 5*s*. 6*d*.

Origenis Philosophumena; sive omnium Haeresium Refutatio. E Codice Parisino nunc primum edidit Emmanuel Miller. 1851. 8vo. *cloth*, 10*s*.

Patrum Apostolicorum, S. Clementis Romani, S. Ignatii, S. Polycarpi, quae supersunt. Edidit Guil. Jacobson, S.T.P.R. Tomi II. *Fourth Edition*, 1863. 8vo. *cloth*, 1*l*. 1*s*.

Reliquiae Sacrae secundi tertiique saeculi. Recensuit M. J. Routh, S.T.P. Tomi V. *Second Edition*, 1846–1848. 8vo. *cloth*, 1*l*. 5*s*.

Scriptorum Ecclesiasticorum Opuscula. Recensuit M. J. Routh, S.T.P. Tomi II. *Third Edition*, 1858. 8vo. *cloth*, 10s.

Socratis Scholastici Historia Ecclesiastica. Gr. et Lat. Edidit R. Hussey, S.T.B. Tomi III. 1853. 8vo. *cloth*, 15s.

Socrates' Ecclesiastical History, according to the Text of Hussey. With an Introduction by William Bright, D.D. Crown 8vo. *cloth*, 7s. 6d. *Just Published.*

Sozomeni Historia Ecclesiastica. Edidit R. Hussey, S.T.B. Tomi III. 1859. 8vo. *cloth*, 15s.

Theodoreti Ecclesiasticae Historiae Libri V. Recensuit T. Gaisford, S.T.P. 1854. 8vo. *cloth*, 7s. 6d.

Theodoreti Graecarum Affectionum Curatio. Ad Codices MSS. recensuit T. Gaisford, S.T.P. 1839. 8vo. *cloth*, 7s. 6d.

Dowling (J. G.) Notitia Scriptorum SS. Patrum aliorumque vet. Eccles. Mon. quae in Collectionibus Anecdotorum post annum Christi MDCC. in lucem editis continentur. 1839. 8vo. *cloth*, 4s. 6d.

ECCLESIASTICAL HISTORY, BIOGRAPHY, &c.

Baedae Historia Ecclesiastica. Edited, with English Notes by G. H. Moberly, M.A. 1869. crown 8vo. *cloth*, 10s. 6d.

Bingham's Antiquities of the Christian Church, and other Works. 10 vols. 1855. 8vo. *cloth*, 3*l*. 3s.

Bright (W., D.D.). Chapters of Early English Church History. 8vo. *cloth*, 12s.

Burnet's History of the Reformation of the Church of England. *A new Edition.* Carefully revised, and the Records collated with the originals, by N. Pocock, M.A. 7 vols. 1865. 8vo. 4*l*. 4s.

Burnet's Life of Sir M. Hale, and Fell's Life of Dr. Hammond. 1856. small 8vo. *cloth*, 2s. 6d.

Cardwell's Two Books of Common Prayer, set forth by authority in the Reign of King Edward VI, compared with each other. *Third Edition*, 1852. 8vo. *cloth*, 7s.

Cardwell's Documentary Annals of the Reformed Church of England; being a Collection of Injunctions, Declarations, Orders, Articles of Inquiry, &c. from 1546 to 1716. 2 vols. 1843. 8vo. *cloth*, 18s.

Cardwell's History of Conferences on the Book of Common Prayer from 1551 to 1690. *Third Edition*, 1849. 8vo. *cloth*, 7s. 6d.

Councils and Ecclesiastical Documents relating to Great Britain and Ireland. Edited, after Spelman and Wilkins, by A. W. Haddan, B.D., and William Stubbs, M.A., Regius Professor of Modern History, Oxford. Vols. I. and III. Medium 8vo. *cloth*, each 1*l*. 1s.

Vol. II. Part I. Medium 8vo. *cloth*, 10s. 6d.

Vol. II. Part II. Church of Ireland; Memorials of St. Patrick. *stiff covers*, 3s. 6d.

Formularies of Faith set forth by the King's Authority during the Reign of Henry VIII. 1856. 8vo. *cloth*, 7*s.*

Fuller's Church History of Britain. Edited by J. S. Brewer, M.A. 6 vols. 1845. 8vo. *cloth*, 1*l.* 19*s.*

Gibson's Synodus Anglicana. Edited by E. Cardwell, D.D. 1854. 8vo. *cloth*, 6*s.*

Hussey's Rise of the Papal Power traced in three Lectures. *Second Edition*, 1863. fcap. 8vo. *cloth*, 4*s.* 6*d.*

Inett's Origines Anglicanae (in continuation of Stillingfleet). Edited by J. Griffiths, M.A. 3 vols. 1855. 8vo. *cloth*, 15*s.*

John, Bishop of Ephesus. The Third Part of his Ecclesiastical History. [In Syriac.] Now first edited by William Cureton, M.A. 1853. 4to. *cloth*, 1*l.* 12*s.*

The same, translated by R. Payne Smith, M.A. 1860. 8vo. *cloth*, 10*s.*

Knight's Life of Dean Colet. 1823. 8vo. *cloth*, 7*s.* 6*d.*

Le Neve's Fasti Ecclesiae Anglicanae. *Corrected and continued from* 1715 *to* 1853 by T. Duffus Hardy. 3 vols. 1854. 8vo. *cloth*, 1*l.* 1*s.*

Noelli (A.) Catechismus sive prima institutio disciplinaque Pietatis Christianae Latine explicata. Editio nova cura Guil. Jacobson, A.M. 1844. 8vo. *cloth*, 5*s.* 6*d.*

Prideaux's Connection of Sacred and Profane History. 2 vols. 1851. 8vo. *cloth*, 10*s.*

Primers put forth in the Reign of Henry VIII. 1848. 8vo. *cloth*, 5*s.*

Records of the Reformation. The Divorce, 1527—1533. Mostly now for the first time printed from MSS. in the British Museum and other Libraries. Collected and arranged by N. Pocock, M.A. 2 vols. 8vo. *cloth*, 1*l.* 16*s.*

Reformatio Legum Ecclesiasticarum. The Reformation of Ecclesiastical Laws, as attempted in the reigns of Henry VIII, Edward VI, and Elizabeth. Edited by E. Cardwell, D.D. 1850. 8vo. *cloth*, 6*s.* 6*d.*

Shirley's (W. W.) Some Account of the Church in the Apostolic Age. *Second Edition*, 1874. fcap. 8vo. *cloth*, 3*s.* 6*d.*

Shuckford's Sacred and Profane History connected (in continuation of Prideaux). 2 vols. 1848. 8vo. *cloth*, 10*s.*

Stillingfleet's Origines Britannicae, with Lloyd's Historical Account of Church Government. Edited by T. P. Pantin, M.A. 2 vols. 1842. 8vo. *cloth*, 10*s.*

Stubbs's (W.) Registrum Sacrum Anglicanum. An attempt to exhibit the course of Episcopal Succession in England. 1858. small 4to. *cloth*, 8*s.* 6*d.*

Strype's Works Complete, with a General Index. 27 vols. 1821–1843. 8vo. *cloth*, 7*l*. 13*s*. 6*d*. Sold separately as follows:—

Memorials of Cranmer. 2 vols. 1840. 8vo. *cloth*, 11*s*.
Life of Parker. 3 vols. 1828. 8vo. *cloth*, 16*s*. 6*d*.
Life of Grindal. 1821. 8vo. *cloth*, 5*s*. 6*d*.
Life of Whitgift. 3 vols. 1822. 8vo. *cloth*, 16*s*. 6*d*.
Life of Aylmer. 1820. 8vo. *cloth*, 5*s*. 6*d*.
Life of Cheke. 1821. 8vo. *cloth*, 5*s*. 6*d*.
Life of Smith. 1820. 8vo. *cloth*, 5*s*. 6*d*.
Ecclesiastical Memorials. 6 vols. 1822. 8vo. *cloth*, 1*l*. 13*s*.
Annals of the Reformation. 7 vols. 8vo. *cloth*, 2*l*. 3*s*. 6*d*.
General Index. 2 vols. 1828. 8vo. *cloth*, 11*s*.

Sylloge Confessionum sub tempus Reformandae Ecclesiae editarum. Subjiciuntur Catechismus Heidelbergensis et Canones Synodi Dordrechtanae. 1827. 8vo. *cloth*, 8*s*.

ENGLISH THEOLOGY.

Beveridge's Discourse upon the XXXIX Articles. *The third complete Edition*, 1847. 8vo. *cloth*, 8*s*.

Bilson on the Perpetual Government of Christ's Church, with a Biographical Notice by R. Eden, M.A. 1842. 8vo. *cloth*, 4*s*.

Biscoe's Boyle Lectures on the Acts of the Apostles. 1840. 8vo. *cloth*, 9*s*. 6*d*.

Bull's Works, with Nelson's Life. By E. Burton, D.D. *A new Edition*, 1846. 8 vols. 8vo. *cloth*, 2*l*. 9*s*.

Burnet's Exposition of the XXXIX Articles. 8vo. *cloth*, 7*s*.

Burton's (Edward) Testimonies of the Ante-Nicene Fathers to the Divinity of Christ. *Second Edition*, 1829. 8vo. *cloth*, 7*s*.

Burton's (Edward) Testimonies of the Ante-Nicene Fathers to the Doctrine of the Trinity and of the Divinity of the Holy Ghost. 1831. 8vo. *cloth*, 3*s*. 6*d*.

Butler's Works, with an Index to the Analogy. 2 vols. 1874. 8vo. *cloth*, 11*s*.

Butler's Sermons. 8vo. *cloth*, 5*s*. 6*d*.

Butler's Analogy of Religion. 8vo. *cloth*, 5*s*. 6*d*.

Chandler's Critical History of the Life of David. 1853. 8vo. *cloth*, 8*s*. 6*d*.

Chillingworth's Works. 3 vols. 1838. 8vo. *cloth*, 1*l*. 1*s*. 6*d*.

Clergyman's Instructor. *Sixth Edition*, 1855. 8vo. *cloth*, 6*s*. 6*d*.

Comber's Companion to the Temple; or a Help to Devotion in the use of the Common Prayer. 7 vols. 1841. 8vo. *cloth*, 1*l*. 11*s*. 6*d*.

Cranmer's Works. Collected and arranged by H. Jenkyns, M.A., Fellow of Oriel College. 4 vols. 1834. 8vo. *cloth*, 1*l*. 10*s*.

Enchiridion Theologicum Anti-Romanum.

Vol. I. Jeremy Taylor's Dissuasive from Popery, and Treatise on the Real Presence. 1852. 8vo. *cloth*, 8s.

Vol. II. Barrow on the Supremacy of the Pope, with his Discourse on the Unity of the Church. 1852. 8vo. *cloth*, 7s. 6d.

Vol. III. Tracts selected from Wake, Patrick, Stillingfleet, Clagett, and others. 1837. 8vo. *cloth*, 11s.

[Fell's] Paraphrase and Annotations on the Epistles of St. Paul. 1852. 8vo. *cloth*, 7s.

Greswell's Harmonia Evangelica. *Fifth Edition*, 1856. 8vo. *cloth*, 9s. 6d.

Greswell's Prolegomena ad Harmoniam Evangelicam. 1840. 8vo. *cloth*, 9s. 6d.

Greswell's Dissertations on the Principles and Arrangement of a Harmony of the Gospels. 5 vols. 1837. 8vo. *cloth*, 3l. 3s.

Hall's (Bp.) Works. *A new Edition*, by Philip Wynter, D.D. 10 vols. 1863. 8vo. *cloth*, 3l. 3s.

Hammond's Paraphrase and Annotations on the New Testament. 4 vols. 1845. 8vo. *cloth*, 1l.

Hammond's Paraphrase on the Book of Psalms. 2 vols. 1850. 8vo. *cloth*, 10s.

Heurtley's Collection of Creeds. 1858. 8vo. *cloth*, 6s. 6d.

Homilies appointed to be read in Churches. Edited by J. Griffiths, M.A. 1859. 8vo. *cloth*, 7s. 6d.

Hooker's Works, with his Life by Walton, arranged by John Keble, M.A. *Sixth Edition*, 1874. 3 vols. 8vo. *cloth*, 1l. 11s. 6d.

Hooker's Works; the text as arranged by John Keble, M.A. 2 vols. 8vo. *cloth*, 11s.

Hooper's (Bp. George) Works. 2 vols. 1855. 8vo. *cloth*, 8s.

Jackson's (Dr. Thomas) Works. 12 vols. 8vo. *cloth*, 3l. 6s.

Jewel's Works. Edited by R. W. Jelf, D.D. 8 vols. 1847. 8vo. *cloth*, 1l. 10s.

Patrick's Theological Works. 9 vols. 1859. 8vo. *cloth*, 1l. 1s.

Pearson's Exposition of the Creed. Revised and corrected by E. Burton, D.D. *Sixth Edition*, 1877. 8vo. *cloth*, 10s. 6d.

Pearson's Minor Theological Works. Now first collected, with a Memoir of the Author, Notes, and Index, by Edward Churton, M.A. 2 vols. 1844. 8vo. *cloth*, 10s.

Sanderson's Works. Edited by W. Jacobson, D.D. 6 vols. 1854. 8vo. *cloth*, 1l. 10s.

Stanhope's Paraphrase and Comment upon the Epistles and Gospels. *A new Edition.* 2 vols. 1851. 8vo. *cloth*, 10s.

Stillingfleet's Origines Sacrae. 2 vols. 1837. 8vo. *cloth*, 9s.

Stillingfleet's Rational Account of the Grounds of Protestant Religion; being a vindication of Abp. Laud's Relation of a Conference, &c. 2 vols. 1844. 8vo. *cloth*, 10s.

Wall's History of Infant Baptism, with Gale's Reflections, and Wall's Defence. *A new Edition*, by Henry Cotton, D.C.L. 2 vols. 1862. 8vo. *cloth*, 1l. 1s.

Waterland's Works, with Life, by Bp. Van Mildert. *A new Edition*, with copious Indexes. 6 vols. 1857. 8vo. *cloth*, 2l. 11s.

Waterland's Review of the Doctrine of the Eucharist, with a Preface by the present Bishop of London. 1868. crown 8vo. *cloth*, 6s. 6d.

Wheatly's Illustration of the Book of Common Prayer. *A new Edition*, 1846. 8vo. *cloth*, 5s.

Wyclif. A Catalogue of the Original Works of John Wyclif, by W. W. Shirley, D.D. 1865. 8vo. *cloth*, 3s. 6d.

Wyclif. Select English Works. By T. Arnold, M.A. 3 vols. 1871. 8vo. *cloth*, 2l. 2s.

Wyclif. Trialogus. *With the Supplement now first edited.* By Gotthardus Lechler. 1869. 8vo. *cloth*, 14s.

ENGLISH HISTORICAL AND DOCUMENTARY WORKS.

British Barrows, a Record of the Examination of Sepulchral Mounds in various parts of England. By William Greenwell, M.A., F.S.A. Together with Description of Figures of Skulls, General Remarks on Prehistoric Crania, and an Appendix. By George Rolleston, M.D., F.R.S. Medium 8vo., *cloth*, 25s.

Two of the Saxon Chronicles parallel, with Supplementary Extracts from the Others. Edited, with Introduction, Notes, and a Glossarial Index, by J. Earle, M.A. 1865. 8vo. *cloth*, 16s.

Magna Carta, a careful Reprint. Edited by W. Stubbs, M.A., Regius Professor of Modern History. 1868. 4to. *stitched*, 1s.

Britton, a Treatise upon the Common Law of England, composed by order of King Edward I. The French Text carefully revised, with an English Translation, Introduction, and Notes, by F. M. Nichols, M.A. 2 vols. 1865. royal 8vo. *cloth*, 1l. 16s.

Burnet's History of His Own Time, with the suppressed Passages and Notes. 6 vols. 1833. 8vo. *cloth*, 2l. 10s.

Burnet's History of James II, with additional Notes. 1852. 8vo. *cloth*, 9s. 6d.

Carte's Life of James Duke of Ormond. *A new Edition*, carefully compared with the original MSS. 6 vols. 1851. 8vo. *cloth*, 1l. 5s.

Casauboni Ephemerides, cum praefatione et notis J. Russell, S.T.P. Tomi II. 1850. 8vo. *cloth*, 15s.

Clarendon's (Edw. Earl of) History of the Rebellion and Civil Wars in England. To which are subjoined the Notes of Bishop Warburton. 7 vols. 1849. medium 8vo. *cloth*, 2*l.* 10*s.*

Clarendon's (Edw. Earl of) History of the Rebellion and Civil Wars in England. 7 vols. 1839. 18mo. *cloth*, 1*l.* 1*s.*

Clarendon's (Edw. Earl of) History of the Rebellion and Civil Wars in England. Also His Life, written by Himself, in which is included a Continuation of his History of the Grand Rebellion. With copious Indexes. In one volume, royal 8vo. 1842. *cloth*, 1*l.* 2*s.*

Clarendon's (Edw. Earl of) Life, including a Continuation of his History. 2 vols. 1857. medium 8vo. *cloth*, 1*l.* 2*s.*

Clarendon's (Edw. Earl of) Life, and Continuation of his History. 3 vols. 1827. 8vo. *cloth*, 16*s.* 6*d.*

Calendar of the Clarendon State Papers, preserved in the Bodleian Library. *In three volumes.*

Vol. I. From 1523 to January 1649. 8vo. *cloth*, 18*s.*
Vol. II. From 1649 to 1654. 8vo. *cloth*, 16*s.*
Vol. III. From 1655 to 1657. 8vo. *cloth*, 14*s.*

Calendar of Charters and Rolls preserved in the Bodleian Library. Edited by W. H. Turner, under the direction of H. O. Coxe, M.A. 8vo. *cloth*, 1*l.* 11*s.* 6*d.* *Just Published.*

Freeman's (E. A.) History of the Norman Conquest of England: its Causes and Results. Vols. I. and II. *Third Edition.* 8vo. *cloth*, 1*l.* 16*s.*

Vol. III. The Reign of Harold and the Interregnum. *Second Edition.* 8vo. *cloth*, 1*l.* 1*s.*
Vol. IV. The Reign of William. *Second Edition.* 8vo. *cloth*, 1*l.* 1*s.*
Vol. V. The Effects of the Norman Conquest. 8vo. *cloth*, 1*l.* 1*s.*

Kennett's Parochial Antiquities. 2 vols. 1818. 4to. *cloth*, 1*l.*

Lloyd's Prices of Corn in Oxford, 1583–1830. 8vo. *sewed*, 1*s.*

Luttrell's (Narcissus) Diary. A Brief Historical Relation of State Affairs, 1678–1714. 6 vols. 1857. 8vo. *cloth*, 1*l.* 4*s.*

May's History of the Long Parliament. 1854. 8vo. *cloth*, 6*s.* 6*d.*

Rogers's History of Agriculture and Prices in England, A.D. 1259–1400. 2 vols. 1866. 8vo. *cloth*, 2*l.* 2*s.*

Sprigg's England's Recovery; being the History of the Army under Sir Thomas Fairfax. *A new edition.* 1854. 8vo. *cloth*, 6*s.*

Whitelock's Memorials of English Affairs from 1625 to 1660. 4 vols. 1853. 8vo. *cloth*, 1*l.* 10*s.*

Protests of the Lords, including those which have been expunged, from 1624 to 1874; with Historical Introductions. Edited by James E. Thorold Rogers, M.A. 3 vols. 8vo. *cloth*, 2*l.* 2*s.*

Enactments in Parliament, specially concerning the Universities of Oxford and Cambridge. Collected and arranged by J. Griffiths, M.A. 1869. 8vo. *cloth*, 12*s*.

Ordinances and Statutes [for Colleges and Halls] framed or approved by the Oxford University Commissioners. 1863. 8vo. *cloth*, 12*s*.—Sold separately (except for Exeter, All Souls, Brasenose, and Corpus), at 1*s*. each.

Statuta Universitatis Oxoniensis. 1878. 8vo. *cloth*, 5*s*.

The Student's Handbook to the University and Colleges of Oxford. *Fourth Edition*. Extra fcap. 8vo. *cloth*, 2*s*. 6*d*.

Index to Wills proved in the Court of the Chancellor of the University of Oxford, &c. Compiled by J. Griffiths, M.A. 1862. royal 8vo. *cloth*, 3*s*. 6*d*.

Catalogue of Oxford Graduates from 1659 to 1850. 1851. 8vo. *cloth*, 7*s*. 6*d*.

CHRONOLOGY, GEOGRAPHY, &c.

Clinton's Fasti Hellenici. The Civil and Literary Chronology of Greece, from the LVIth to the CXXIIIrd Olympiad. *Third edition*, 1841. 4to. *cloth*, 1*l*. 14*s*. 6*d*.

Clinton's Fasti Hellenici. The Civil and Literary Chronology of Greece, from the CXXIVth Olympiad to the Death of Augustus. *Second edition*, 1851. 4to. *cloth*, 1*l*. 12*s*.

Clinton's Epitome of the Fasti Hellenici. 8vo. *cloth*, 6*s*. 6*d*.

Clinton's Fasti Romani. The Civil and Literary Chronology of Rome and Constantinople, from the Death of Augustus to the Death of Heraclius. 2 vols. 1845, 1850. 4to. *cloth*, 3*l*. 9*s*.

Clinton's Epitome of the Fasti Romani. 1854. 8vo. *cloth*, 7*s*.

Cramer's Geographical and Historical Description of Asia Minor. 2 vols. 1832. 8vo. *cloth*, 11*s*.

Cramer's Map of Asia Minor, 15*s*.

Cramer's Map of Ancient and Modern Italy, on two sheets, 15*s*.

Cramer's Description of Ancient Greece. 3 vols. 1828. 8vo. *cloth*, 16*s*. 6*d*.

Cramer's Map of Ancient and Modern Greece, on two sheets, 15*s*.

Greswell's Fasti Temporis Catholici. 4 vols. 8vo. *cloth*, 2*l*. 10*s*.

Greswell's Tables to Fasti, 4to., and Introduction to Tables, 8vo. *cloth*, 15*s*.

Greswell's Origines Kalendariæ Italicæ. 4 vols. 8vo. *cloth*, 2*l*. 2*s*.

Greswell's Origines Kalendariæ Hellenicæ. 6 vols. 1862. 8vo. *cloth*, 4*l*. 4*s*.

PHILOSOPHICAL WORKS, AND GENERAL LITERATURE.

The Logic of Hegel; translated from the Encyclopaedia of the Philosophical Sciences. With Prolegomena. By William Wallace, M.A. 8vo. *cloth*, 14s.

Bacon's Novum Organum. Edited, with English notes, by G. W. Kitchin, M.A. 1855. 8vo. *cloth*, 9s. 6d

Bacon's Novum Organum. Translated by G. W. Kitchin, M.A. 1855. 8vo. *cloth*, 9s. 6d. (See also p. 31.)

The Works of George Berkeley, D.D., formerly Bishop of Cloyne; including many of his writings hitherto unpublished. With Prefaces, Annotations, and an Account of his Life and Philosophy, by Alexander Campbell Fraser, M.A. 4 vols. 1871. 8vo. *cloth*, 2*l*. 18s.

The Life, Letters, &c. 1 vol. *cloth*, 16s. *See also* p. 31.

Smith's Wealth of Nations. A new Edition, with Notes, by J. E. Thorold Rogers, M.A. 2 vols. 1870. *cloth*, 21s.

A Course of Lectures on Art, delivered before the University of Oxford in Hilary Term, 1870. By John Ruskin, M.A., Slade Professor of Fine Art. 8vo. *cloth*, 6s.

A Critical Account of the Drawings by Michel Angelo and Raffaello in the University Galleries, Oxford. By J. C. Robinson, F.S.A. Crown 8vo. *cloth*, 4s.

MATHEMATICS, PHYSICAL SCIENCE, &c.

Archimedis quae supersunt omnia cum Eutocii commentariis ex recensione Josephi Torelli, cum novâ versione Latinâ. 1792. folio. *cloth*, 1*l*. 5s.

Bradley's Miscellaneous Works and Correspondence. With an Account of Harriot's Astronomical Papers. 1832. 4to. *cloth*, 17s.

Reduction of Bradley's Observations by Dr. Busch. 1838. 4to. *cloth*, 3s.

Astronomical Observations made at the University Observatory, Oxford, under the direction of C. Pritchard, M. A. No. 1. Royal 8vo. *paper covers*, 3s. 6d.

A Treatise on the Kinetic Theory of Gases. By Henry William Watson, M.A., formerly Fellow of Trinity College, Cambridge. 1876. 8vo. *cloth*, 3s. 6d.

Rigaud's Correspondence of Scientific Men of the 17th Century, with Table of Contents by A. de Morgan, and Index by the Rev. J. Rigaud, M.A., Fellow of Magdalen College, Oxford. 2 vols. 1841–1862. 8vo. *cloth*, 18s. 6d.

Treatise on Infinitesimal Calculus. By Bartholomew Price, M.A., F.R.S., Professor of Natural Philosophy, Oxford.

Vol. I. Differential Calculus. *Second Edition*, 8vo. *cloth*, 14s. 6d.

Vol. II. Integral Calculus, Calculus of Variations, and Differential Equations. *Second Edition*, 1865. 8vo. *cloth*, 18s.

Vol. III. Statics, including Attractions; Dynamics of a Material Particle. *Second Edition*, 1868. 8vo. *cloth*, 16s.

Vol. IV. Dynamics of Material Systems; together with a Chapter on Theoretical Dynamics, by W. F. Donkin, M.A., F.R.S. 1862. 8vo. *cloth*, 16s.

Daubeny's Introduction to the Atomic Theory. 16mo. *cloth*, 6s.

Vesuvius. By John Phillips, M.A., F.R.S., Professor of Geology, Oxford. 1869. Crown 8vo. *cloth*, 10s. 6d.

Geology of Oxford and the Valley of the Thames. By the same Author. 8vo. *cloth*, 21s.

Synopsis of the Pathological Series in the Oxford Museum. By H. W. Acland, M.D., F.R.S., 1867. 8vo. *cloth*, 2s. 6d.

Thesaurus Entomologicus Hopeianus, or a Description of the rarest Insects in the Collection given to the University by the Rev. William Hope. By J. O. Westwood, M.A. With 40 Plates, mostly coloured. Small folio, *half morocco*, 7l. 10s.

Text-Book of Botany, Morphological and Physiological. By Dr. Julius Sachs, Professor of Botany in the University of Würzburg. Translated by A. W. Bennett, M.A., assisted by W. T. Thiselton Dyer, M.A. Royal 8vo. *half morocco*, 1l. 11s. 6d.

On Certain Variations in the Vocal Organs of the Passeres that have hitherto escaped notice. By Johannes Muller. Translated by F. J. Bell, B.A. With an Appendix by A. H. Garrod, M.A., F.R.S. With Plates. 4to. *paper covers*, 7s. 6d.

BIBLIOGRAPHY.

Ebert's Bibliographical Dictionary, translated from the German. 4 vols. 1837. 8vo. *cloth*, 1l. 10s.

Cotton's List of Editions of the Bible in English. *Second Edition*, corrected and enlarged. 1852. 8vo. *cloth*, 8s. 6d.

Cotton's Typographical Gazetteer. 1831. 8vo. *cloth*, 12s. 6d.

Cotton's Typographical Gazetteer, Second Series. 1866. 8vo. *cloth*, 12s. 6d.

Cotton's Rhemes and Doway. An attempt to shew what has been done by Roman Catholics for the diffusion of the Holy Scriptures in English. 1855. 8vo. *cloth*, 9s.

Clarendon Press Series.

The Delegates of the Clarendon Press having undertaken the publication of a series of works, chiefly educational, and entitled the Clarendon Press Series, have published, or have in preparation, the following.

Those to which prices are attached are already published; the others are in preparation.

I. ENGLISH.

A First Reading Book. By Marie Eichens of Berlin; and edited by Anne J. Clough. Extra fcap. 8vo. *stiff covers*, 4*d.*

Oxford Reading Book, Part I. For Little Children. Extra fcap. 8vo. *stiff covers*, 6*d.*

Oxford Reading Book, Part II. For Junior Classes. Extra fcap. 8vo. *stiff covers*, 6*d.*

An Elementary English Grammar and Exercise Book. By O. W. Tancock, M.A., Assistant Master of Sherborne School. Extra fcap. 8vo. *cloth*, 1*s.* 6*d.*

An English Grammar and Reading Book, for Lower Forms in Classical Schools. By O. W. Tancock, M.A., Assistant Master of Sherborne School. *Third Edition.* Extra fcap. 8vo. *cloth*, 3*s.* 6*d.*

Typical Selections from the best English Writers, with Introductory Notices. *Second Edition.* In Two Volumes. Extra fcap. 8vo. *cloth*, 3*s.* 6*d.* each.

Vol. I. Latimer to Berkeley. Vol. II. Pope to Macaulay.

The Philology of the English Tongue. By J. Earle, M.A., formerly Fellow of Oriel College, and sometime Professor of Anglo-Saxon, Oxford. *Second Edition.* Extra fcap. 8vo. *cloth*, 7*s.* 6*d.*

A Book for the Beginner in Anglo-Saxon. By John Earle, M.A., Professor of Anglo-Saxon, Oxford. Extra fcap. 8vo. *cloth*, 2*s.* 6*d.*

An Anglo-Saxon Reader. In Prose and Verse. With Grammatical Introduction, Notes, and Glossary. By Henry Sweet, M.A. Extra fcap. 8vo. *cloth*, 8*s.* 6*d.*

The Ormulum; with the Notes and Glossary of Dr. R. M. White. Edited by Rev. R. Holt, M.A. 2 vols. Extra fcap. 8vo. *cloth*, 21*s.* *Just Published.*

Specimens of Early English. A New and Revised Edition. With Introduction, Notes, and Glossarial Index. By R. Morris, LL.D., and W. W. Skeat, M.A.

Part I. *In the Press.*

Part II. From Robert of Gloucester to Gower (A.D. 1298 to A.D. 1393). *Second Edition.* Extra fcap. 8vo. *cloth*, 7*s.* 6*d.*

Specimens of English Literature, from the 'Ploughmans Crede' to the 'Shepheardes Calender' (A.D. 1394 to A.D. 1579). With Introduction, Notes, and Glossarial Index. By W. W. Skeat, M.A. Extra fcap. 8vo. *cloth*, 7*s.* 6*d.*

The Vision of William concerning Piers the Plowman, by William Langland. Edited, with Notes, by W. W. Skeat, M.A. *Second Edition.* Extra fcap. 8vo. *cloth*, 4*s.* 6*d.*

Chaucer. The Prioresses Tale; Sir Thopas; The Monkes Tale; The Clerkes Tale; The Squieres Tale, &c. Edited by W. W. Skeat, M.A. *Second Edition.* Extra fcap. 8vo. *cloth*, 4*s.* 6*d.*

Chaucer. The Tale of the Man of Lawe; The Pardoneres Tale; The Second Nonnes Tale; The Chanouns Yemannes Tale. By the same Editor. Extra fcap. 8vo. *cloth*, 4*s.* 6*d.* (See also p. 20.)

Old English Drama. Marlowe's Tragical History of Dr. Faustus, and **Greene's** Honourable History of Friar Bacon and Friar Bungay. Edited by A. W. Ward, M.A., Professor of History and English Literature in Owens College, Manchester. Extra fcap. 8vo. *cloth*, 5*s.* 6*d.*

Shakespeare. Hamlet. Edited by W. G. Clark, M.A., and W. Aldis Wright, M.A. Extra fcap. 8vo. *stiff covers*, 2*s.*

Shakespeare. Select Plays. Edited by W. Aldis Wright, M.A. Extra fcap. 8vo. *stiff covers.*

The Tempest, 1*s.* 6*d.*	King Lear, 1*s.* 6*d.*
As You Like It, 1*s.* 6*d.*	A Midsummer Night's Dream, 1*s.* 6*d.*
Julius Cæsar, 2*s.*	Coriolanus. *In the Press.*

(For other Plays, see p. 20.)

Milton. Areopagitica. With Introduction and Notes. By J. W. Hales, M.A., late Fellow of Christ's College, Cambridge. *Second Edition.* Extra fcap. 8vo. *cloth*, 3*s.*

Addison. Selections from Papers in the Spectator. With Notes. By T. Arnold, M.A., University College. *Second Edition.* Extra fcap. 8vo. *cloth*, 4*s.* 6*d.*

Burke. Four Letters on the Proposals for Peace with the Regicide Directory of France. Edited, with Introduction and Notes, by E. J. Payne, M.A. Extra fcap. 8vo. *cloth*, 5*s.* (See also p. 21.)

Also the following in paper covers :—

Gray. Elegy, and Ode on Eton College. 2*d.*

Johnson. Vanity of Human Wishes. With Notes by E. J. Payne, M.A. 4*d.*

Keats. Hyperion, Book I. With Notes by W. T. Arnold, B.A. 4*d.*

Milton. With Notes by R. C. Browne, M.A.

Lycidas, 3*d.* L'Allegro, 3*d.* Il Penseroso, 4*d.* Comus, 6*d.* Samson Agonistes, 6*d.*

Parnell. The Hermit. 2*d.*

A SERIES OF ENGLISH CLASSICS,

Designed to meet the wants of Students in English Literature, under the superintendence of the Rev. J. S. Brewer, M.A., *of Queen's College, Oxford, and Professor of English Literature at King's College, London.*

It is also especially hoped that this Series may prove useful to Ladies' Schools and Middle Class Schools; in which English Literature must always be a leading subject of instruction.

A General Introduction to the Series. By Professor Brewer, M.A.

1. **Chaucer.** The Prologue to the Canterbury Tales; The Knightes Tale; The Nonne Prestes Tale. Edited by R. Morris, Editor of Specimens of Early English, &c., &c. *Sixth Edition.* Extra fcap. 8vo. *cloth*, 2*s.* 6*d.* (See also p. 19.)

2. **Spenser's Faery Queene.** Books I and II. Designed chiefly for the use of Schools. With Introduction, Notes, and Glossary. By G. W. Kitchin, M.A., formerly Censor of Christ Church.

   Book I. *Eighth Edition.* Extra fcap. 8vo. *cloth*, 2*s.* 6*d.*

   Book II. *Third Edition.* Extra fcap. 8vo. *cloth*, 2*s.* 6*d.*

3. **Hooker.** Ecclesiastical Polity, Book I. Edited by R. W. Church, M.A., Dean of St. Paul's; formerly Fellow of Oriel College, Oxford. *Second Edition.* Extra fcap. 8vo. *cloth*, 2*s.*

4. **Shakespeare.** Select Plays. Edited by W. G. Clark, M.A., Fellow of Trinity College, Cambridge; and W. Aldis Wright, M.A., Trinity College, Cambridge. Extra fcap. 8vo. *stiff covers.*

   I. The Merchant of Venice. 1*s.*

   II. Richard the Second. 1*s.* 6*d.*

   III. Macbeth. 1*s.* 6*d.* (For other Plays, see p. 19.)

5. **Bacon.**

I. Advancement of Learning. Edited by W. Aldis Wright, M.A. *Second Edition.* Extra fcap. 8vo, *cloth*, 4*s*. 6*d*.

II. The Essays. With Introduction and Notes. By J. R. Thursfield, M.A., Fellow and formerly Tutor of Jesus College, Oxford.

6. **Milton.** Poems. Edited by R. C. Browne, M.A. 2 vols. *Fourth Edition.* Extra fcap. 8vo. *cloth*, 6*s*. 6*d*.

Sold separately, Vol. I. 4*s*.; Vol. II. 3*s*. (See also pp. 19, 20.)

7. **Dryden.** Select Poems. Stanzas on the Death of Oliver Cromwell; Astræa Redux; Annus Mirabilis; Absalom and Achitophel; Religio Laici; The Hind and the Panther. Edited by W. D. Christie, M.A. *Second Edition.* Ext. fcap. 8vo. *cloth*, 3*s*. 6*d*.

8. **Bunyan.** The Pilgrim's Progress; Grace Abounding. Edited by E. Venables, M.A., Canon of Lincoln. *Nearly Ready.*

9. **Pope.** With Introduction and Notes. By Mark Pattison, B.D., Rector of Lincoln College, Oxford.

I. Essay on Man. *Fifth Edition.* Extra fcap. 8vo. 1*s*. 6*d*.

II. Satires and Epistles. *Second Edition.* Extra fcap. 8vo. 2*s*.

10. **Johnson.** Rasselas; Lives of Pope and Dryden. Edited by Alfred Milnes, B.A. (London), late Scholar of Lincoln College, Oxford. *In the Press.*

11. **Burke.** Select Works. Edited, with Introduction and Notes, by E. J. Payne, M.A., of Lincoln's Inn, Barrister-at-Law, and Fellow of University College, Oxford.

I. Thoughts on the Present Discontents; the two Speeches on America. *Second Edition.* Extra fcap. 8vo. *cloth*, 4*s*. 6*d*.

II. Reflections on the French Revolution. *Second Edition.* Extra fcap. 8vo. *cloth*, 5*s*. (See also p. 19.)

12. **Cowper.** Edited, with Life, Introductions, and Notes, by H. T. Griffith, B.A., formerly Scholar of Pembroke College, Oxford.

I. The Didactic Poems of 1782, with Selections from the Minor Pieces, A.D. 1779–1783. Extra fcap. 8vo. *cloth*, 3*s*.

II. The Task, with Tirocinium, and Selections from the Minor Poems, A.D. 1784–1799. Extra fcap. 8vo. *cloth*, 3*s*.

II. LATIN.

An Elementary Latin Grammar. By John B. Allen, M.A., Head Master of Perse Grammar School, Cambridge. *Second Edition, Revised and Corrected.* Extra fcap. 8vo. *cloth*, 2*s*. 6*d*.

A First Latin Exercise Book. By the same Author. *Second Edition.* Extra fcap. 8vo. *cloth*, 2*s*. 6*d*.

A Series of Graduated Latin Readers.

First Latin Reader. By T. J. Nunns, M.A. *Third Edition.* Extra fcap. 8vo. *cloth,* 2s.

Second Latin Reader. *In Preparation.*

Third Latin Reader, or Specimens of Latin Literature. Part I, Poetry. By James McCall Marshall, M.A., Dulwich College.

Fourth Latin Reader.

Cicero. Selection of interesting and descriptive passages. With Notes. By Henry Walford, M.A. In three Parts. *Second Edition.* Extra fcap. 8vo. *cloth,* 4s. 6*d.*

Each Part separately, limp, 1s. 6*d.*

Part I. Anecdotes from Grecian and Roman History.
Part II. Omens and Dreams: Beauties of Nature.
Part III. Rome's Rule of her Provinces.

Cicero. Selected Letters (for Schools). With Notes. By the late C. E. Prichard, M.A., and E. R. Bernard, M.A. *Second Edition.* Extra fcap. 8vo. *cloth,* 3s.

Pliny. Selected Letters (for Schools). With Notes. By the same Editors. *Second Edition.* Extra fcap. 8vo. *cloth,* 3s.

Cornelius Nepos. With Notes. By Oscar Browning, M.A. *Second Edition.* Extra fcap. 8vo. *cloth,* 2s. 6*d.*

Caesar. The Commentaries (for Schools). With Notes and Maps. By Charles E. Moberly, M.A.

Part I. The Gallic War. *Third Edition.* Extra fcap. 8vo. *cloth,* 4s. 6*d.*
Part II. The Civil War. Extra fcap. 8vo. *cloth,* 3s. 6*d.*
The Civil War. Book I. Extra fcap. 8vo. *cloth,* 2s.

Livy. Selections (for Schools). With Notes and Maps. By H. Lee-Warner, M.A., Assistant Master in Rugby School. Extra fcap. 8vo. *In Parts, limp, each* 1s. 6*d.*

Part I. The Caudine Disaster.
Part II. Hannibal's Campaign in Italy.
Part III. The Macedonian War.

Livy, Books I–X. By J. R. Seeley, M.A., Regius Professor of Modern History, Cambridge. Book I. *Second Edition.* 8vo. *cloth,* 6s.

Also a small edition for Schools.

Luciani Vera Historia. With Introduction and Notes. By C. S. Jerram, M.A. *Just ready.*

Tacitus. The Annals. Books I–VI. With Essays and Notes. By T. F. Dallin, M.A., Tutor of Queen's College, Oxford. *Preparing.*

Passages for Translation into Latin. For the use of Passmen and others. Selected by J. Y. Sargent, M.A., Fellow and Tutor of Magdalen College, Oxford. *Fifth Edition.* Ext. fcap. 8vo. *cloth,* 2s. 6*d.*

Cicero's Philippic Orations. With Notes. By J. R. King, M.A., formerly Fellow and Tutor of Merton College. *Second Edition.* 8vo. *cloth,* 10s. 6d.

Cicero. Select Letters. With English Introductions, Notes, and Appendices. By Albert Watson, M.A., Fellow and formerly Tutor of Brasenose College, Oxford. *Second Edition.* Demy 8vo. *cloth,* 18s.

Cicero. Select Letters. *Text.* By the same Editor. Extra fcap. 8vo. *cloth,* 4s.

Cicero pro Cluentio. With Introduction and Notes. By W. Ramsay, M.A. Edited by G. G. Ramsay, M.A., Professor of Humanity, Glasgow. Extra fcap. 8vo. *cloth,* 3s. 6d.

Cicero de Oratore. Book I. With Introduction and Notes. By A. S. Wilkins, M.A., Professor of Latin, Owens College, Manchester. *Just ready.*

Catulli Veronensis Liber. Iterum recognovit, apparatum criticum prolegomena appendices addidit, Robinson Ellis, A.M. Demy 8vo. *cloth,* 16s.

A Commentary on Catullus. By Robinson Ellis, M.A. Demy 8vo. *cloth,* 16s.

Catulli Veronensis Carmina Selecta, secundum recognitionem Robinson Ellis, A.M. Extra fcap. 8vo. *cloth,* 3s. 6d.

Horace. With a Commentary. Volume I. The Odes, Carmen Seculare, and Epodes. By Edward C. Wickham, M.A., Head Master of Wellington College. *Second Edition.* 8vo. *cloth,* 12s.

Also a small edition for Schools.

Ovid. Selections for the use of Schools. With Introductions and Notes, and an Appendix on the Roman Calendar. By W. Ramsay, M.A. Edited by G. G. Ramsay, M.A., Professor of Humanity, Glasgow. *Second Edition.* Ext. fcap. 8vo. *cloth,* 5s. 6d.

Persius. The Satires. With a Translation and Commentary. By John Conington, M.A. Edited by Henry Nettleship, M.A. *Second Edition.* 8vo. *cloth,* 7s. 6d.

Selections from the less known Latin Poets. By North Pinder, M.A. Demy 8vo. *cloth,* 15s.

Fragments and Specimens of Early Latin. With Introductions and Notes. By John Wordsworth, M.A. 8vo. *cloth,* 18s.

Vergil: Suggestions Introductory to a Study of the Aeneid. By H. Nettleship, M.A. 8vo. *sewed,* 1s. 6d.

The Roman Satura: its original form in connection with its literary development. By H. Nettleship, M.A. 8vo. *sewed,* 1s.

A Manual of Comparative Philology. By T. L. Papillon, M.A., Fellow and Lecturer of New College. *Second Edition.* Crown 8vo. *cloth*, 6s.

The Roman Poets of the Augustan Age. By William Young Sellar, M.A., Professor of Humanity in the University of Edinburgh. VIRGIL. 8vo. *cloth*, 14s.

The Roman Poets of the Republic. By the same Editor. *Preparing.*

III. GREEK.

A Greek Primer, for the use of beginners in that Language. By the Right Rev. Charles Wordsworth, D.C.L., Bishop of St. Andrews. *Sixth Edition, Revised and Enlarged.* Extra fcap. 8vo. *cloth*, 1s. 6d.

Graecae Grammaticae Rudimenta in usum Scholarum. Auctore Carolo Wordsworth, D.C.L. *Eighteenth Edition*, 1875. 12mo. *bound*, 4s.

A Greek-English Lexicon, abridged from Liddell and Scott's 4to. edition, chiefly for the use of Schools. *Seventeenth Edition. Carefully Revised throughout.* 1876. Square 12mo. *cloth*, 7s. 6d.

Greek Verbs, Irregular and Defective; their forms, meaning, and quantity; embracing all the Tenses used by Greek writers, with reference to the passages in which they are found. By W. Veitch. *New Edition.* Crown 8vo. *cloth*, 10s. 6d.

The Elements of Greek Accentuation (for Schools): abridged from his larger work by H. W. Chandler, M.A., Waynflete Professor of Moral and Metaphysical Philosophy, Oxford. Ext. fcap. 8vo. *cloth*, 2s. 6d.

A Series of Graduated Greek Readers.

First Greek Reader. By W. G. Rushbrooke, M.L. Extra fcap. 8vo. *cloth*, 2s. 6d.

Second Greek Reader. By A. J. M. Bell, M.A. *In the Press.*

Third Greek Reader. *In Preparation.*

Fourth Greek Reader; being Specimens of Greek Dialects. With Introductions and Notes. By W. W. Merry, M.A., Fellow and Lecturer of Lincoln College. Extra fcap. 8vo. *cloth*, 4s. 6d.

Fifth Greek Reader. Part I. Selections from Greek Epic and Dramatic Poetry, with Introductions and Notes. By Evelyn Abbott, M.A., Fellow of Balliol College. Ext. fcap. 8vo. *cloth*, 4s. 6d.

Part II. By the same Editor. *In Preparation.*

Aeschylus. Prometheus Bound (for Schools). With Introduction and Notes, by A. O. Prickard, M.A., Fellow of New College. Extra fcap. 8vo. *cloth*, 2s.

Aristophanes. Nubes (for Schools). With Introduction, Notes, etc. By W. W. Merry, M.A. *Preparing.*

Xenophon. Anabasis, Book II. With Notes and Map. By C. S. Jerram, M.A. Extra fcap. 8vo. *cloth*, 2s.

Xenophon. Easy Selections (for Junior Classes). With a Vocabulary, Notes, and Map. By J. S. Phillpotts, B.C.L., and C. S. Jerram, M.A. Extra fcap. 8vo. *cloth*, 3s. 6*d*.

Xenophon. Selections (for Schools). With Notes and Maps. By J. S. Phillpotts, B.C.L., Head Master of Bedford School. *Fourth Edition*. Ext. fcap. 8vo. *cloth*, 3s. 6*d*.

Arrian. Selections (for Schools). With Notes. By J. S. Phillpotts, B.C.L., Head Master of Bedford School.

Cebes. Tabula. With Introduction and Notes. By C. S. Jerram, M.A. Extra fcap. 8vo. *cloth*, 2s. 6*d*.

The Golden Treasury of Ancient Greek Poetry; being a Collection of the finest passages in the Greek Classic Poets, with Introductory Notices and Notes. By R. S. Wright, M.A., Fellow of Oriel College, Oxford. Ext. fcap. 8vo. *cloth*, 8s. 6*d*.

A Golden Treasury of Greek Prose, being a collection of the finest passages in the principal Greek Prose Writers, with Introductory Notices and Notes. By R. S. Wright, M.A., and J. E. L. Shadwell, M.A. Ext. fcap. 8vo. *cloth*, 4s. 6*d*.

Aristotle's Politics. By W. L. Newman, M.A., Fellow of Balliol College, Oxford.

Demosthenes and Aeschines. The Orations of Demosthenes and Æschines on the Crown. With Introductory Essays and Notes. By G. A. Simcox, M.A., and W. H. Simcox, M.A. 8vo. *cloth*, 12s.

Theocritus (for Schools). With Notes. By H. Kynaston, M.A. (late Snow), Head Master of Cheltenham College. *Second Edition*. Extra fcap. 8vo. *cloth*, 4s. 6*d*.

Homer. Odyssey, Books I—XII (for Schools). By W. W. Merry, M.A. *Fifth Edition*. Extra fcap. 8vo. *cloth*, 4s. 6*d*.

Book II, *separately*, 1s. 6*d*.

Homer. Odyssey, Books XIII-XXIV (for Schools). By the same Editor. Extra fcap. 8vo. *cloth*, 5s.

Homer. Odyssey, Books I-XII. Edited with English Notes, Appendices, etc. By W. W. Merry, M.A., and the late James Riddell, M.A. Demy 8vo. *cloth*, 16s.

Homer. Odyssey, Books XIII-XXIV. With Introduction and Notes. By S. H. Butcher, M.A., Fellow of University College.

Homer. Iliad, Book I (for Schools). By D. B. Monro, M.A. Extra fcap. 8vo. *cloth*, 2s.

Homer. Iliad. With Introduction and Notes. By D. B. Monro, M.A. *Preparing*.

A Homeric Grammar. By D. B. Monro, M.A. *Preparing*.

Plato. Selections (for Schools). With Notes. By B. Jowett, M.A., Regius Professor of Greek; and J. Purves, M.A., Fellow and Lecturer of Balliol College, Oxford. *In the Press.*

Sophocles. The Plays and Fragments. With English Notes and Introductions. By Lewis Campbell, M.A., Professor of Greek, St. Andrews, formerly Fellow of Queen's College, Oxford. 2 vols.

Vol. I. Oedipus Tyrannus. Oedipus Coloneus. Antigone. 8vo. *cloth*, 14*s.*

Sophocles. The Text of the Seven Plays. By the same Editor. Ext. fcap. 8vo. *cloth*, 4*s.* 6*d.*

Sophocles. In Single Plays, with English Notes, &c. By Lewis Campbell, M.A., and Evelyn Abbott, M.A. Extra fcap. 8vo. *limp.*

Oedipus Rex, Oedipus Coloneus, Antigone, 1*s.* 9*d.* each.
Ajax, Electra, Trachiniae, 2*s.* each.

Sophocles. Oedipus Rex: Dindorf's Text, with Notes by the present Bishop of St. David's. Ext. fcap. 8vo. *limp*, 1*s.* 6*d.*

IV. FRENCH.

An Etymological Dictionary of the French Language, with a Preface on the Principles of French Etymology. By A. Brachet. Translated into English by G. W. Kitchin, M.A., formerly Censor of Christ Church. *Second Edition.* Crown 8vo. *cloth. Price reduced to* 7*s.* 6*d.*

Brachet's Historical Grammar of the French Language. Translated into English by G. W. Kitchin, M.A. *Fourth Edition.* Extra fcap. 8vo. *cloth*, 3*s.* 6*d.*

French Classics, Edited by Gustave Masson, B.A.

Corneille's Cinna, and **Molière's** Les Femmes Savantes. With Introduction and Notes. Extra fcap. 8vo. *cloth*, 2*s.* 6*d.*

Racine's Andromaque, and **Corneille's** Le Menteur. With Louis Racine's Life of his Father. Extra fcap. 8vo. *cloth*, 2*s.* 6*d.*

Molière's Les Fourberies de Scapin, and **Racine's** Athalie. With Voltaire's Life of Molière. Extra fcap. 8vo. *cloth*, 2*s.* 6*d.*

Selections from the Correspondence of **Madame de Sévigné** and her chief Contemporaries. Intended more especially for Girls' Schools. Extra fcap. 8vo. *cloth*, 3*s.*

Voyage autour de ma Chambre, by **Xavier de Maistre**; Ourika, by **Madame de Duras**; La Dot de Suzette, by **Fievée**; Les Jumeaux de l'Hôtel Corneille, by **Edmond About**; Mésaventures d'un Écolier, by **Rodolphe Töpffer.** Extra fcap. 8vo. *cloth*, 2*s.* 6*d.*

Regnard's Le Joueur, and **Brueys** and **Palaprat's** Le Grondeur. Extra fcap. 8vo. *cloth*, 2*s.* 6*d.*

Louis XIV and his Contemporaries; as described in Extracts from the best Memoirs of the Seventeenth Century. With English Notes, Genealogical Tables, &c. By the same Editor. Extra fcap. 8vo. *cloth*, 2s. 6*d*.

V. GERMAN.

LANGE'S *German Course. By* HERMANN LANGE, *Teacher of Modern Languages, Manchester:*

The Germans at Home; a Practical Introduction to German Conversation, with an Appendix containing the Essentials of German Grammar. *Second Edition.* 8vo. *cloth*, 2s. 6*d*.

The German Manual; a German Grammar, a Reading Book, and a Handbook of German Conversation. 8vo. *cloth*, 7s. 6*d*.

A Grammar of the German Language. 8vo. *cloth*, 3*s*. 6*d*.

This 'Grammar' is a reprint of the Grammar contained in 'The German Manual,' and, in this separate form, is intended for the use of students who wish to make themselves acquainted with German Grammar chiefly for the purpose of being able to read German books.

German Composition; Extracts from English and American writers for Translation into German, with Hints for Translation in foot-notes. *In the Press.*

Lessing's Laokoon. With Introduction, English Notes, etc. By A. HAMANN, Phil. Doc., M.A., Taylorian Teacher of German in the University of Oxford. Extra fcap. 8vo. *cloth*, 4s. 6*d*.

Goethe's Faust. Part I. With Introduction and Notes. By the same Editor. *In Preparation.*

Wilhelm Tell. A Drama. By Schiller. Translated into English Verse by E. Massie, M.A. Extra fcap. 8vo. *cloth*, 5s.

Also, *Edited by* C. A. BUCHHEIM, *Phil. Doc., Professor in King's College, London.*

Goethe's Egmont. With a Life of Goethe, &c. *Second Edition.* Extra fcap. 8vo. *cloth*, 3s.

Schiller's Wilhelm Tell. With a Life of Schiller; an historical and critical Introduction, Arguments, and a complete Commentary. *Third Edition.* Extra fcap. 8vo. *cloth*, 3s. 6*d*.

Lessing's Minna von Barnhelm. A Comedy. With a Life of Lessing, Critical Analysis, Complete Commentary, &c. *Second Edition.* Extra fcap. 8vo. *cloth*, 3s. 6*d*.

Schiller's Egmonts Leben und Tod, and Belagerung von Antwerpen. Extra fcap. 8vo. *cloth*, 2s. 6*d*.

In Preparation.

Goethe's Iphigenie auf Tauris. A Drama. With a Critical Introduction, Arguments to the Acts, and a complete Commentary.

Selections from the Poems of **Schiller** and **Goethe**.

Becker's (K. F.) Friedrich der Grosse.

VI. MATHEMATICS, &c.

Figures Made Easy: a first Arithmetic Book. (Introductory to 'The Scholar's Arithmetic.') By Lewis Hensley, M.A., formerly Fellow and Assistant Tutor of Trinity College, Cambridge. Crown 8vo. *cloth*, 6*d*.

Answers to the Examples in Figures made Easy, together with two thousand additional Examples formed from the Tables in the same, with Answers. By the same Author. Crown 8vo. *cloth*, 1*s*.

The Scholar's Arithmetic; with Answers to the Examples. By the same Author. Crown 8vo. *cloth*, 4*s*. 6*d*.

The Scholar's Algebra. An Introductory work on Algebra. By the same Author. Crown 8vo. *cloth*, 4*s*. 6*d*.

Book-keeping. By R. G. C. Hamilton, Financial Assistant Secretary to the Board of Trade, and John Ball (of the Firm of Quilter, Ball, & Co.), Co-Examiners in Book-keeping for the Society of Arts. *New and enlarged Edition.* Extra fcap. 8vo. *limp cloth*, 2*s*.

A Course of Lectures on Pure Geometry. By Henry J. Stephen Smith, M.A., F.R.S., Fellow of Corpus Christi College, and Savilian Professor of Geometry in the University of Oxford.

Acoustics. By W. F. Donkin, M.A., F.R.S., Savilian Professor of Astronomy, Oxford. Crown 8vo. *cloth*, 7*s*. 6*d*.

A Treatise on Electricity and Magnetism. By J. Clerk Maxwell, M.A., F.R.S., Professor of Experimental Physics in the University of Cambridge. 2 vols. 8vo. *cloth*, 1*l*. 11*s*. 6*d*.

An Elementary Treatise on the same subject. By the same Author. *Preparing.*

VII. PHYSICAL SCIENCE.

A Handbook of Descriptive Astronomy. By G. F. Chambers, F.R.A.S., Barrister-at-Law. *Third Edition.* Demy 8vo. *cloth*, 28*s*.

Chemistry for Students. By A. W. Williamson, Phil Doc., F.R.S., Professor of Chemistry, University College, London. *A new Edition, with Solutions.* Extra fcap. 8vo. *cloth*, 8*s*. 6*d*.

A Treatise on Heat, with numerous Woodcuts and Diagrams. By Balfour Stewart, LL.D., F.R.S., Professor of Natural Philosophy in Owens College, Manchester. *Third Edition.* Extra fcap. 8vo. *cloth*, 7*s*. 6*d*.

Lessons on Thermodynamics. By R. E. Baynes, M.A., Senior Student of Christ Church, Oxford, and Lee's Reader in Physics. Crown 8vo. *cloth*, 7*s*. 6*d*.

Forms of Animal Life. By G. Rolleston, M.D., F.R.S., Linacre Professor of Physiology, Oxford. Illustrated by Descriptions and Drawings of Dissections. Demy 8vo. *cloth*, 16s.

Exercises in Practical Chemistry (Laboratory Practice). By A. G. Vernon Harcourt, M.A., F.R.S., Senior Student of Christ Church, and Lee's Reader in Chemistry; and H. G. Madan, M.A., Fellow of Queen's College, Oxford. *Second Edition.* Crown 8vo. *cloth.* 7s. 6*d*.

Geology of Oxford and the Valley of the Thames. By John Phillips, M.A., F.R.S., Professor of Geology, Oxford. 8vo. *cloth*, 21s.

Crystallography. By M. H. N. Story-Maskelyne, M.A., Professor of Mineralogy, Oxford; and Deputy Keeper in the Department of Minerals, British Museum. *In the Press.*

VIII. HISTORY.

The Constitutional History of England, in its Origin and Development. By William Stubbs, M.A., Regius Professor of Modern History. *In Three Volumes.* Crown 8vo. *cloth*, each 12s.

Select Charters and other Illustrations of English Constitutional History, from the Earliest Times to the Reign of Edward I. Arranged and Edited by W. Stubbs, M.A. *Third Edition.* Crown 8vo. *cloth*, 8s. 6*d*.

A History of England, principally in the Seventeenth Century. By Leopold Von Ranke. Translated by Resident Members of the University of Oxford, under the superintendence of G. W. Kitchin, M.A., and C. W. Boase, M.A. 6 vols. 8vo. *cloth*, 3*l*. 3*s*.

Genealogical Tables illustrative of Modern History. By H. B. George, M.A. *Second Edition.* Small 4to. *cloth*, 12*s*.

A History of France. With numerous Maps, Plans, and Tables. By G. W. Kitchin, M.A. *In Three Volumes.* Crown 8vo. *cloth*, each 10s. 6*d*.

Vol. 1. Down to the Year 1453. Vol. 2. From 1453–1624.
Vol. 3. From 1624–1793.

A Manual of Ancient History. By George Rawlinson, M.A., Camden Professor of Ancient History, formerly Fellow of Exeter College, Oxford. Demy 8vo. *cloth*, 14s.

A History of Germany and of the Empire, down to the close of the Middle Ages. By J. Bryce, D.C.L., Regius Professor of Civil Law in the University of Oxford.

A History of British India. By S. J. Owen, M.A., Reader in Indian History in the University of Oxford.

A History of Greece. By E. A. Freeman, M.A., formerly Fellow of Trinity College, Oxford.

A History of Greece from its Conquest by the Romans to the present time, B.C. 146 to A.D. 1864. By George Finlay, LL.D. A new Edition, revised throughout, and in part re-written, with considerable additions, by the Author, and Edited by H. F. Tozer, M.A., Tutor and late Fellow of Exeter College, Oxford. In Seven Volumes. 8vo. *cloth*, 3*l*. 10*s*.

A Selection from the Despatches, Treaties, and other Papers of the Marquess Wellesley, K.G., during his Government of India; with Appendix, Map of India, and Plans. Edited by S. J. Owen, M.A., Reader in Indian History in the University of Oxford, formerly Professor of History in the Elphinstone College, Bombay. 8vo. *cloth*, 1*l*. 4*s*.

IX. LAW.

Elements of Law considered with reference to Principles of General Jurisprudence. By William Markby, M.A., Judge of the High Court of Judicature, Calcutta. *Second Edition, with Supplement.* Crown 8vo. *cloth*, 7*s*. 6*d*.

An Introduction to the History of the Law of Real Property, with original Authorities. By Kenelm E. Digby, M.A., of Lincoln's Inn, Barrister-at-Law, and formerly Fellow of Corpus Christi College, Oxford. *Second Edition.* Crown 8vo. *cloth*, 7*s*. 6*d*.

The Elements of Jurisprudence. By Thomas Erskine Holland, D.C.L., Chichele Professor of International Law and Diplomacy, and formerly Fellow of Exeter College, Oxford. *In the Press.*

The Institutes of Justinian, edited as a recension of the Institutes of Gaius. By the same Editor. Extra fcap. 8vo. *cloth*, 5*s*.

Alberici Gentilis, I.C.D., I.C. Professoris Regii, De Iure Belli Libri Tres. Edidit Thomas Erskine Holland I.C.D., Iuris Gentium Professor Chicheleianus, Coll. Omn. Anim. Socius, necnon in Univ. Perusin. Iuris Professor Honorarius. Small 4to. *half morocco*, 21*s*.

Gaii Institutionum Juris Civilis Commentarii Quatuor; or, Elements of Roman Law by Gaius. With a Translation and Commentary by Edward Poste, M.A., Barrister-at-Law, and Fellow of Oriel College, Oxford. *Second Edition.* 8vo. *cloth*, 18*s*.

Select Titles from the Digest of Justinian. By T. E. Holland, D.C.L., Chichele Professor of International Law and Diplomacy, and formerly Fellow of Exeter College, Oxford, and C. L. Shadwell, B.C.L., Fellow of Oriel College, Oxford. *In Parts.*

Part I. Introductory Titles. 8vo. *sewed*, 2*s*. 6*d*.
Part II. Family Law. 8vo. *sewed*, 1*s*.
Part III. Property Law. 8vo. *sewed*, 2*s*. 6*d*.
Part IV. (No. 1). **Law of Obligations.** 8vo. *sewed*, 3*s*. 6*d*.

Principles of the English Law of Contract. By Sir William R. Anson, Bart., B.C.L., Vinerian Reader of English Law, and Fellow of All Souls College, Oxford. *In the Press.*

X. MENTAL AND MORAL PHILOSOPHY.

Bacon. Novum Organum. Edited, with Introduction, Notes, &c., by T. Fowler, M.A., Professor of Logic in the University of Oxford. 8vo. *cloth*, 14s.

Selections from Berkeley, with an Introduction and Notes. For the use of Students in the Universities. By Alexander Campbell Fraser, LL.D. Crown 8vo. *cloth*, 7s. 6d. *See also* p. 16.

The Elements of Deductive Logic, designed mainly for the use of Junior Students in the Universities. By T. Fowler, M.A., Professor of Logic in the University of Oxford. *Sixth Edition*, with a Collection of Examples. Extra fcap. 8vo. *cloth*, 3s. 6d.

The Elements of Inductive Logic, designed mainly for the use of Students in the Universities. By the same Author. *Third Edition*. Extra fcap. 8vo. *cloth*, 6s.

A Manual of Political Economy, for the use of Schools. By J. E. Thorold Rogers, M.A., formerly Professor of Political Economy, Oxford. *Third Edition*. Extra fcap. 8vo. *cloth*, 4s. 6d.

An Introduction to the Principles of Morals and Legislation. By Jeremy Bentham. Crown 8vo. *cloth*, 6s. 6d.

XI. ART, &c.

A Handbook of Pictorial Art. By R. St. J. Tyrwhitt, M.A., formerly Student and Tutor of Christ Church, Oxford. With coloured Illustrations, Photographs, and a chapter on Perspective by A. Macdonald. *Second Edition*. 8vo. *half morocco*, 18s.

A Music Primer for Schools. By J. Troutbeck, M.A., Music Master in Westminster School, and R. F. Dale, M.A., B. Mus., Assistant Master in Westminster School. Crown 8vo. *cloth*, 1s. 6d.

A Treatise on Harmony. By Sir F. A. Gore Ouseley, Bart., Professor of Music in the University of Oxford. *Second Edition*. 4to. *cloth*, 10s.

A Treatise on Counterpoint, Canon, and Fugue, based upon that of Cherubini. By the same Author. 4to. *cloth*, 16s.

A Treatise on Musical Form and General Composition. By the same Author. 4to. *cloth*, 10s.

The Cultivation of the Speaking Voice. By John Hullah. *Second Edition*. Extra fcap. 8vo. *cloth*, 2s. 6d.

XII. MISCELLANEOUS.

Specimens of Lowland Scotch and Northern English. By Dr. J. A. H. Murray. *Preparing.*

Dante. Selections from the Inferno. With Introduction and Notes. By H. B. Cotterill, B.A. Extra fcap. 8vo. *cloth*, 4*s*. 6*d*.

Tasso. La Gerusalemme Liberata. Cantos i, ii. With Introduction and Notes. By the same Editor. Extra fcap. 8vo. *cloth*, 2*s*. 6*d*.

A Treatise on the use of the Tenses in Hebrew. By S. R. Driver, M.A., Fellow of New College. Extra fcap. 8vo. *cloth*, 6*s*. 6*d*.

The Book of Tobit. A Chaldee Text, from a unique MS. in the Bodleian Library; with other Rabbinical Texts, English Translations, and the Itala. Edited by Ad. Neubauer, M.A. Crown 8vo. *cloth*, 6*s*.

Outlines of Textual Criticism applied to the New Testament. By C. E. Hammond, M.A., Fellow and Tutor of Exeter College, Oxford. *Second Edition.* Extra fcap. 8vo. *cloth*, 3*s*. 6*d*.

The Modern Greek Language in its relation to Ancient Greek. By E. M. Geldart, B.A. Extra fcap. 8vo. *cloth*, 4*s*. 6*d*.

A Handbook of Phonetics, including a Popular Exposition of the Principles of Spelling Reform. By Henry Sweet, President of the Philological Society, Author of a 'History of English Sounds,' &c. Extra fcap. 8vo. *cloth*, 4*s*. 6*d*.

A System of Physical Education: Theoretical and Practical. By Archibald Maclaren. Extra fcap. 8vo. *cloth*, 7*s*. 6*d*.

Published for the University by
MACMILLAN AND CO., LONDON.

Also to be had at the
CLARENDON PRESS DEPOSITORY, OXFORD.

The DELEGATES OF THE PRESS *invite suggestions and advice from all persons interested in education; and will be thankful for hints, &c. addressed to the* SECRETARY TO THE DELEGATES, *Clarendon Press, Oxford.*

www.ingramcontent.com/pod-product-compliance
Lightning Source LLC
Chambersburg PA
CBHW021932190326
41519CB00009B/996
* 9 7 8 3 7 4 2 8 9 2 2 8 7 *